Kousik Das Malakar

Environmental impact of the COVID-19 pandemic: A lesson for the future

AF536753

Kousik Das Malakar

Environmental impact of the COVID-19 pandemic: A lesson for the future

Coronavirus, Consequences, Physical, Economic, Socio-cultural, and Political perspective

JustFiction Edition

Imprint
Any brand names and product names mentioned in this book are subject to trademark, brand or patent protection and are trademarks or registered trademarks of their respective holders. The use of brand names, product names, common names, trade names, product descriptions etc. even without a particular marking in this work is in no way to be construed to mean that such names may be regarded as unrestricted in respect of trademark and brand protection legislation and could thus be used by anyone.

Cover image: www.ingimage.com

Publisher:
JustFiction! Edition
is a trademark of
International Book Market Service Ltd., member of OmniScriptum Publishing Group
17 Meldrum Street, Beau Bassin 71504, Mauritius
Printed at: see last page
ISBN: 978-620-0-49468-9

Copyright © Kousik Das Malakar
Copyright © 2020 International Book Market Service Ltd., member of OmniScriptum Publishing Group

Environmental impact of the COVID-19 pandemic: A lesson for the future

(Coronavirus, Consequences, Physical, Economic, Socio-cultural, and Political perspective)

Kousik Das Malakar

Jawaharlal Nehru University (JNU), New Delhi, India

Just Fiction! Edition

Founded in Germany

Some other books by the same Author

- The Upper-Caste Community of Malakar and their hereditary livelihoods
- Scope and Contents of Digital Humanities
- Population and Environment: A Review (Part-6)
- Population and Environment: A Review (Part-5)
- Population and Environment: A Review (Part-4)
- Population and Environment: A Review (Part-3)
- Population and Environment: A Review (Part-2)
- Population and Environment: A Review (Part-1)
- A Regional Study of Bankura District in West Bengal, India.
- Digital humanities of India.
- Physical Characteristics of Baragere Village, Kharagpur-2, W.B.
- Socio-Economic Conditions of Baragere Village, Medinipur West, West Bengal.
- Agricultural Conditions of Baragere Village, Medinipur West, West Bengal.
- The Mangrove: Future of the Global seacoast.
- Project Field Report: Rural Development of India (Physical and Socio-Economic Conditions of Baragere Village, Kharagpur-II, Paschim Medinipur, West Bengal).
- A Basic Outline of Geomorphic Studies (Part-2).
- A Basic Outline of Geomorphic Studies (Part-1).
- Coronavirus (COVID-19): A Pandemic Situation in the World.
- Geospatial Technology: An Overview Concept of RS, GIS, and GPS.
- A Basic Outline of Population Studies.
- Climate Change: A Thinkable Manifesto of Future.
- Climatic Data Analysis and Regional Specialty (Part - II).
- Climatic Data Analysis and Regional Specialty (Part - I).
- Regional Identity of India (Part - 2).
- Regional Identity of India (Part – I).
- Agricultural Infrastructure and Land Use in India.
- River Basin Hydrological Measurement Technique.
- Social Studies: A Study of Public Health in India.
- Regional Studies: Some thinkable subjective discussion.

Dedications

My dear Parents

PREFACE

This book is an outcome of present days pandemic Situation in the World. As a student of geography, I observed that the vulnerable situation and try to get a basic configuration of COVID-19 worldwide.

The book, ***'Environmental impact of the COVID-19 pandemic: A lesson for the future'*** is comprehensive and easy to understand the text. This book discussed a general introduction of COVID-19, symptoms, causes, diagnosis, prevention, management, research, affected countries and their related reports (till 23rd March of 2020), history of the coronavirus pandemic, responses and socio-political consequences of COVID-19, and three case studies included, which is express the situations of Physical, Economic, Socio-cultural, and Political perspective.

At least, I humbly hope that this study will contribute to the increasing public awareness and perceptions regarding COVID-19. Suggestions for improving the book are always welcome and will be incorporated in the next edition.

JNU, New Delhi. **Kousik Das Malakar**
September 4th, 2020

ACKNOWLEDGEMENT

As I was finalizing the manuscript of this book, my memory goes back to identify all those Physical and human geographers who positively affected me throughout my career and in many cases, they were not aware of the impact they were having on me. They range from my graduate Professors at Jawaharlal Nehru University (JNU) and Vidyasagar University (VU) in my early days to present geographers, Professors, colleagues, friends all over the country. To all of them, I express my regard and gratitude.

In writing this book, I have got immense help from a vast range of publications, So, I thank all the authors of the publications.

And I also grateful to my parents, because they are good wishes to me.

JNU, New Delhi. **Kousik Das Malakar**

September 4th, 2020

CONTENTS

Chapter – 1

Introduction

Concept:

The term of 'epidemic', 'epi' means above and 'demos' means people (from Greek word). An epidemic is the rapid spread of disease to a large number of people in a given population within a short period of time. For example, in meningococcal infections, an attack rate in excess of 15 cases per 100,000 people for two consecutive weeks is considered an epidemic.

The term epidemic derives from a word form attributed to Homer's Odyssey, which later took its medical meaning from the Epidemics, a treatise by Hippocrates. Before Hippocrates, epidemios, epidemeo, epidamos, and other variants had meanings similar to the current definitions of "indigenous" or "endemic". Thucydides' description of the Plague of Athens is considered one of the earliest accounts of a disease epidemic. By the early 17th century, the terms endemic and epidemic referred to contrasting conditions of population-level disease, with the endemic condition at low rates of occurrence and the epidemic condition widespread. The term "epidemic" has become emotionally charged. The Atlanta Centre for Disease Control defines epidemic broadly: "Epidemic: the occurrence of more cases of disease, injury, or other health condition than expected in a given area or among a specific group of persons during a particular period. Usually, the cases are presumed to have a common cause or to be related to one another in some way. The terms "epidemic" and "outbreak" have often been used interchangeably. Researchers Manfred S. Green and colleagues propose that the latter term be restricted to smaller events, pointing out that Chambers Concise Dictionary and Stedman's Medical Dictionary acknowledge this distinction.

Epidemics of infectious disease are generally caused by several factors including a change in the ecology of the host population (e.g. increased stress or increase in the density of a vector species), a genetic change in the pathogen reservoir or the introduction of an emerging pathogen to a host population (by movement of pathogen or host). Generally, an epidemic occurs when host immunity to either an established pathogen or newly emerging novel pathogen is suddenly reduced below that found in the endemic equilibrium and the transmission threshold is exceeded.

An epidemic may be restricted to one location; however, if it spreads to other countries or continents and affects a substantial number of people, it may be termed a pandemic. The declaration of an epidemic usually requires a good understanding of a baseline rate of incidence; epidemics for certain diseases, such as COVID-19, influenza, are defined as reaching some defined increase in incidence above this baseline. A few cases of a very rare disease may be classified as an epidemic, while many cases of a common disease (such as the common cold) would not.

Causes of the Epidemic:

There are several changes that may occur in an infectious agent that may trigger an epidemic. These include; Increased virulence, Introduction into a novel setting, Changes in host susceptibility to the infectious agent. An epidemic disease is not required to be contagious, and the term has been applied to West Nile fever and the obesity epidemic (e.g. by the World Health Organisation), among others.

The conditions which govern the outbreak of epidemics include infected food supplies such as contaminated drinking water and the migration of populations of certain animals, such as rats or mosquitoes, which can act as disease vectors. Certain epidemics occur at certain seasons. For example, whooping-cough occurs in spring, whereas measles produces two epidemics, one in winter and one in March. Influenza, the common cold, and other infections of the upper respiratory tract, such as sore throat, occur predominantly in the winter. There is another variation, both as regards the number of people affected and the number who die in successive epidemics: the severity of successive epidemics rises and falls over periods of five or ten years.

Various types:

Common source outbreak

In a common source outbreak epidemic, the affected individuals had an exposure to a common agent. If the exposure is singular and all of the affected individuals develop the disease over a single exposure and incubation course, it can be termed a point source outbreak. If the exposure was continuous or variable, it can be termed a continuous outbreak or intermittent outbreak, respectively.

Propagated outbreak

In a propagated outbreak, the disease spreads person-to-person. Affected individuals may become independent reservoirs leading to further exposures. Many epidemics will have characteristics of both common source and propagated outbreaks (sometimes referred as mixed outbreak). For example, secondary person-to-person spread may occur after a common source exposure or an environmental vectors may spread a zoonotic diseases agent.

Transmission:

- **Airborne transmission:** Airborne transmission is the spread of infection by droplet nuclei or dust in the air. Without the intervention of winds or drafts the distance over which airborne infection takes place is short, say 10 to 20 feet.
- **Arthropod transmission:** Arthropod transmission takes place by an insect, either mechanically through a contaminated proboscis or feet, or biologically when there is growth or replication of an organism in the arthropod.
- **Biological transmission:** Involving a biological process, e.g. passing a stage of development of the infecting agent in an intermediate host. Opposite to mechanical transmission.
- **Contact transmission:** The disease agent is transferred directly by biting, sucking, chewing or indirectly by inhalation of droplets, drinking of contaminated water, traveling in contaminated vehicles.
- **Cyclopropagative transmission:** The agent undergoes both development and multiplication in the transmitting vehicle.
- **Developmental transmission:** The agent undergoes some development in the transmission vehicle.
- **Fecal-oral transmission:** The infectious agent is shed by the infected host in feces and acquired by the susceptible host through ingestion of contaminated material.
- **Horizontal transmission:** Lateral spread to others in the same group and at the same time; spread to contemporaries.

- **Propagative transmission:** The agent multiplies in the transmission vehicle.
- **Vertical transmission:** From one generation to the next, perhaps transovarially or by intrauterine infection of the fetus. Some retroviruses are transmitted in the germ line, i.e. their genetic material is integrated into the DNA of either the ovum or sperm.

Preparation

Experts suggest that the best way to prepare for an epidemic is to have a disease surveillance system, be able to quickly dispatch emergency workers, especially local-based emergency workers, and have a legitimate way to guarantee the safety and health of health workers. Effective preparations for a response to a pandemic are multi-layered. The first layer is a disease surveillance system. Tanzania, for example, runs a national lab that runs testing for 200 health sites and tracks the spread of infectious diseases. The next layer is the actual response to an emergency. According to U.S.-based columnist Michael Gerson, only the U.S. military and NATO have the global capability to respond to such an emergency.

Century-wise list of epidemics in the World:

15th Century & earlier				
Death toll (estimate)	**Location**	**Date**	**Event**	**Disease**
75,000–100,000	Greece	429–426 BC	Plague of Athens	Unknown, possibly typhus, typhoid fever or viral hemorrhagic fever
Unknown	Greece (Northern Greece, Roman Republic)	412 BC	412 BC epidemic	Unknown, possibly influenza
5–10 million	Roman Empire	165–180 (possibly up to 190)	Antonine Plague	Unknown, possibly smallpox

1 million + (Unknown, but at least)	Europe	250–266	Plague of Cyprian	Unknown, possibly smallpox
100 million (Up to); 40–50% of population of Europe	Europe and West Asia	541–542	Plague of Justinian	Plague
	British Isles	664–689	Plague of 664	Plague
2 million (Approx. ⅓ of entire Japanese population)	Japan	735–737	735–737 Japanese smallpox epidemic	Smallpox
	Byzantine Empire, West Asia, Africa	746–747	Plague of 746–747	Plague
50–200 million (10–60% of European population)	Europe, Asia and North Africa	1331–1353	Black Death	Plague
10,000 +	Britain (England) and later continental Euro pe	1485–1551	Sweating sickness (m ultiple outbreaks)	Unknown, possibly an unknown species of hantavirus

16th and 17th centuries				
Death toll (estimate)	**Location**	**Date**	**Event**	**Disease**
5–8 million	Mexico	1520	1520 Smallpox Epidemic	Smallpox

5–15 million (80% of population)	Mexico	1545–1548	Cocoliztli Epidemic of 1545–1548	Possibly Salmonella enterica
> 20,100 in London	London	1563–1564	1563 London plague	Plague
2–2.5 million (50% of population)	Mexico	1576–1580	Cocoliztli epidemic of 1576	Possibly Salmonella enterica
	Seneca nation	1592–1596		Measles
3,000	Malta	1592–1593	1592–93 Malta plague epidemic	Plague
> 19,900 in London and outer parishes	London	1592–1593	1592–93 London plague	Plague
	Spain	1596–1602		Plague
	South America	1600–1650		Malaria
	England	1603		Plague
	Egypt	1609		Plague

30–90% of population	Southern New England, especially the Wampanoag people	1616–1620	1616 New England epidemic	Unknown cause. Latest research suggests epidemic(s) of leptospirosis with W eil syndrome. Classic explanations include yellow fever, bubonic plague, influenza, smal lpox, chickenpox, typh us, and syndemic infection of hepatitis B and hepatitis D.
2,80,000	Italy	1629–1631	Italian plague of 1629–1631	Plague
	Wyandot people	1634		Smallpox
	Thirteen Colonies	1633	Massachusetts smallpox epidemic	Smallpox
	England	1636		Plague
	China	1641–1644		Plague
	Spain	1647–1652	Great Plague of Seville	Plague
	Central America	1648		Yellow fever
	Italy	1656	Naples Plague	Plague
	Thirteen Colonies	1657		Measles
24,148	Netherlands	1663–1664		Plague

100,000	England	1665–1666	Great Plague of London	Plague
40,000	France	1668		Plague
11,300	Malta	1675–1676	1675–76 Malta plague epidemic	Plague
	Spain	1676–1685		Plague
76,000	Austria	1679	Great Plague of Vienna	Plague
	Thirteen Colonies	1687		Measles
	Thirteen Colonies	1690		Yellow fever

18th century				
Death toll (estimate)	**Location**	**Date**	**Event**	**Disease**
	Canada, New France	1702–1703		Smallpox
> 18,000 (36% of population)	Iceland	1707–1709	Great Smallpox Epidemic	Smallpox
	Denmark, Sweden	1710–1712	Great Northern War plague outbreak	Plague
	Thirteen Colonies	1713–1715		Measles
	Canada, New France	1714–1715		Measles
>100,000	France	1720–1722	Great Plague of Marseille	Plague
	Thirteen Colonies	1721–1722		Smallpox
	Thirteen Colonies	1729		Measles
	Spain	1730		Yellow fever
	Thirteen Colonies	1732–1733		Influenza

	Canada, New France	1733		Smallpox
> 50,000	Balkans	1738	Great Plague of 1738	Plague
	Thirteen Colonies	1738		Smallpox
	Thirteen Colonies	1739–1740		Measles
	Italy	1743		Plague
	Thirteen Colonies	1747		Measles
	North America	1755–1756		Smallpox
	North America	1759		Measles
	North America, West Indies	1761		Influenza
	North America, present-day Pittsburgh area.	1763		Smallpox
> 50,000	Russia	1770–1772	Russian plague of 1770–1772	Plague
	Pacific Northwest natives	1770s		Smallpox
	North America	1772		Measles
> 2,000,000	Persia	1772	Persian Plague	Plague
	England	1775–1776		Influenza
	Spain	1778		Dengue fever
	Plains Indians	1780–1782	North American smallpox epidemic	Smallpox
	Pueblo Indians	1788		Smallpox
	United States	1788		Measles

	New South Wales, Australia	1789–1790		Smallpox
	United States	1793		Influenza and epidemic typhus
	United States	1793–1798	Yellow Fever Epidemic of 1793, resurgences	Yellow fever

19th century				
Death toll (estimate)	**Location**	**Date**	**Event**	**Disease**
	Spain	1800–1803		Yellow fever
	Ottoman Empire, Egypt	1801		Bubonic plague
	United States	1803		Yellow fever
	Egypt	1812		Plague
	Ottoman Empire	1812–19	1812–19 Ottoman plague epidemic	Plague
4,500	Malta	1813–1814	1813–14 Malta plague epidemic	Plague
60,000	Romania	1813	Caragea's plague	Plague
	Ireland	1816–1819		Typhus
> 100,000	Asia, Europe	1816–1826	First cholera pandemic	Cholera
	United States	1820–1823		Yellow fever
	Spain	1821		Yellow fever

	New South Wales, Australia	1828		Smallpox
	Netherlands	1829	Groningen epidemic	Malaria
	South Australia	1829		Smallpox
	Iran	1829–1835		Bubonic plague
> 100,000	Asia, Europe, North America	1829–1851	Second cholera pandemic	Cholera
	Egypt	1831		Cholera
	Plains Indians	1831–1834		Smallpox
	England, France	1832		Cholera
	North America	1832		Cholera
	United States	1833		Cholera
	United States	1834		Cholera
	Egypt	1834–1836		Bubonic plague
	United States	1837		Typhus
	Great Plains	1837–1838	1837–38 smallpox epidemic	Smallpox
	Dalmatia	1840		Plague
	South Africa	1840		Smallpox
	United States	1841		Yellow fever
> 20,000	Canada	1847–1848	Typhus epidemic of 1847	Epidemic typhus
	United States	1847		Yellow fever
	Worldwide	1847–1848		Influenza
	Egypt	1848		Cholera
	North America	1848–1849		Cholera

	United States	1850		Yellow fever
	North America	1850–1851		Influenza
	United States	1851		Cholera
	United States	1852		Yellow fever
10,00,000	Russia	1852–1860	Third cholera pandemic	Cholera
	Ottoman Empire	1853		Plague
4,737	Copenhagen, Denmark	1853	Cholera epidemic of Copenhagen 1853	Cholera
616	England	1854	Broad Street cholera outbreak	Cholera
	United States	1855		Yellow fever
>22 million in India, more worldwide	Worldwide	1855–1960	Third plague pandemic	Bubonic plague
	Portugal	1857		Yellow fever
	Victoria, Australia	1857		Smallpox
	Europe, North America, South America	1857–1859		Influenza
> 3,000	Central Coast, British Columbia	1862–1863		Smallpox
	Middle East	1863–1879	Fourth cholera pandemic	Cholera
	Egypt	1865		Cholera
	Russia, Germany	1866–1867		Cholera

	Australia	1867		Measles
	Iraq	1867		Plague
	Argentina	1852–1871		Yellow fever
	Germany	1870–1871		Smallpox
40,000	Fiji	1875	1875 Fiji Measles outbreak	Measles
	Russian Empire	1877		Plague
	Egypt	1881		Cholera
> 9,000	India, Germany	1881–1896	Fifth cholera pandemic	Cholera
3,164	Montreal	1885		Smallpox
10,00,000	Worldwide	1889–1890	1889–1890 flu pandemic	Influenza
	West Africa	1900		Yellow fever

20th Century				
Death toll (estimate)	**Location**	**Date**	**Event**	**Disease**
	Congo Basin	1896–1906		Trypanosomiasis
> 800,000	Europe, Asia, Africa	1899–1923	Sixth cholera pandemic	Cholera
113	San Francisco	1900–1904		Bubonic plague
	Uganda	1900–1920		Trypanosomiasis
	Egypt	1902		Cholera
22	India	1903		Bubonic Plague
4	Fremantle	1903		Bubonic plague
40,000	China	1910–1912	1910 China plague	Bubonic plague
1.5 million	worldwide	1915–1926	1915 Encephalitis lethargica pandemic	Encephalitis lethargica

up to 100,000,000	worldwide	1918–1920	Spanish flu (pandemic)	Influenza A virus subtype H1N1
	Russia	1918–1922		Typhus
30	Los Angeles	1924	1924 Los Angeles pneumonic plague outbreak	Pneumonic plague
43	Croydon, UK	1937	Croydon epidemic of typhoid fever	Typhoid fever
	Egypt	1942–1944		Malaria
	China	1946		Bubonic plague
	Egypt	1946		Relapsing fever
	Egypt	1947		Cholera
20,00,000	worldwide	1957–1958	Asian flu	Influenza A virus subtype H2N2
	worldwide	1961–1975	Seventh cholera pandemic	Cholera (El Tor strain)
500 million between 1877–1977	worldwide	From 3rd century BCE until its eradication in 1977		Smallpox
10,00,000	worldwide	1968–1969	Hong Kong flu	Influenza A virus subtype H3N2
5	Netherlands	1971		Poliomyelitis
35	Yugoslavia	1972	1972 outbreak of smallpox in Yugoslavia	Smallpox

1,027	United States	1972–1973	London flu	Influenza A virus subtype H3N2
15,000	India	1974	1974 smallpox epidemic of India	Smallpox
> 32,000,000	worldwide	1920–present	HIV/AIDS pandemic	HIV/AIDS
	South America	1990s		Cholera
52	India	1994	1994 plague epidemic in Surat	Plague
231	worldwide	1996–2001		vCJD
	West Africa	1996		Meningitis
105	Malaysia	1998–1999	1998–99 Malaysia Nipah virus outbreak	Nipah virus infection
	Central America	2000		Dengue fever

21st Century				
Death toll (estimate)	**Location**	**Date**	**Event**	**Disease**
> 400	Nigeria	2001		Cholera
	South Africa	2001		Cholera
359	China	2002–2004	2002–2004 SARS outbreak	Severe acute respiratory syndrome (SARS)
299	Hong Kong			
37	Taiwan			
44	Canada			
33	Singapore			
	Algeria	2003		Plague
	Afghanistan	2004		Leishmaniasis
	Bangladesh	2004		Cholera
	Indonesia	2004		Dengue fever

	Senegal	2004		Cholera
7	Sudan	2004		Ebola
	Mali	2005		Yellow fever
27	Singapore	2005	2005 dengue outbreak in Singapore	Dengue fever
	Luanda, Angola	2006		Cholera
61	Ituri Province, Democratic Republic of the Congo	2006		Plague
17	India	2006		Malaria
> 50	India	2006	2006 dengue outbreak in India	Dengue fever
	India	2006	Chikungunya outbreaks	Chikungunya virus
> 50	Pakistan	2006	2006 dengue outbreak in Pakistan	Dengue fever
	Philippines	2006		Dengue fever
187	Democratic Republic of the Congo	2007	Mweka ebola epidemic	Ebola
	Ethiopia	2007		Cholera
49	India	2008		Cholera
10	Iraq	2007	2007 Iraq cholera outbreak	Cholera
	Nigeria	2007		Poliomyelitis

	Puerto Rico; Dominican Republic; Mexico	2007		Dengue fever
	Somalia	2007		Cholera
37	Uganda	2007		Ebola
	Vietnam	2007		Cholera
	Brazil	2008		Dengue fever
	Cambodia	2008		Dengue fever
	Chad	2008		Cholera
	China	2008–2017		Hand, foot and mouth disease
	Madagascar	2008		Bubonic plague
	Philippines	2008		Dengue fever
	Vietnam	2008		Cholera
4,293	Zimbabwe	2008–2009	2008–2009 Zimbabwean cholera outbreak	Cholera
18	Bolivia	2009	2009 Bolivian dengue fever epidemic	Dengue fever
49	India	2009	2009 Gujarat hepatitis outbreak	Hepatitis B
	Queensland, Australia	2009		Dengue fever
	Worldwide	2009	Mumps outbreaks in the 2000s	Mumps

931	West Africa	2009–2010	2009–2010 West African meningitis outbreak	Meningitis
151,700 – 575,400	Worldwide	2009	2009 flu pandemic (informally called "swine flu")	Pandemic H1N1/09 virus
10,075 (May 2017)	Hispaniola	2010–present	Haiti cholera outbreak	Cholera (strain serogroup O1, serotype Ogawa)
> 4,500 (February 2014)	Democratic Republic of the Congo	2011–present		Measles
170	Vietnam	2011–present		Hand, foot and mouth disease
> 350	Pakistan	2011	2011 dengue outbreak in Pakistan	Dengue fever
171 (as of 10 January 2013)	Darfur Sudan	2012	2012 yellow fever outbreak in Darfur, Sudan	Yellow fever
862 (as of 13 January 2020)	Worldwide	2012–present	2012 Middle East respiratory syndrome coronavirus outbreak	Middle East respiratory syndrome (MERS)

142	Vietnam	2013–2014		Measles
>> 11,300	Worldwide, primarily concentrated in West Africa	2013–2016	Ebola virus epidemic in West Africa	Ebola virus disease
183	Americas	2013–2015	2013–14 chikungunya outbreak	Chikungunya
40	Madagascar	2014–2017	2014 Madagascar plague outbreak	Bubonic plague
36	India	2014–2015	2014 Odisha jaundice outbreak	Primarily Hepatitis E, but also Hepatitis A
2,035	India	2015	2015 Indian swine flu outbreak	Influenza A virus subtype H1N1
~53	Worldwide	2015–2016	2015–16 Zika virus epidemic	Zika virus
100s (as of 1 April 2016)	Africa	2016	2016 yellow fever outbreak in Angola	Yellow fever
3,886 (as of 30 November 2019)	Yemen	2016–present	2016–20 Yemen cholera outbreak	Cholera
64 (as of 16 August 2017)	India	2017	2017 Gorakhpur Japanese encephalitis outbreak	Japanese encephalitis

18 (as of February 2020)	India	2018	2018 Nipah virus outbreak in Kerala	Nipah virus infection
2,253 (as of 20 February 2020)	Democratic Republic of the Congo & Uganda	August 2018–present	2018–20 Kivu Ebola epidemic	Ebola virus disease
>5,000 (by November 2019)	Democratic Republic of the Congo	2019–present	2019 measles outbreak in the Democratic Republic of the Congo	Measles
83	Samoa	2019–present	2019 Samoa measles outbreak	Measles
2,000 (Approx)	Asia-Pacific, Latin America	2019–present	2019 dengue fever epidemic	Dengue fever
17,151+ **(as of 23 March 2020)**	**Worldwide**	**2019–present**	**2019–20 COVID-19 pandemic**	**Coronavirus disease 2019 (COVID-19)**

****

Chapter- 2

A General introduction of COVID-19

Introduction:

A coronavirus is a virus that is found in animals and, rarely, can be transmitted from animals to humans and then spread person to person. In addition to COVID-19, other human coronaviruses have included; the MERS virus, or Middle East respiratory syndrome; the SARS virus, or severe acute respiratory syndrome, which first occurred in the Guangdong province in southern China.

It was discovered in December 2019 and has now spread throughout the world. As the virus spreads, we are seeing some people with mild illness, some who get very sick, and some who have died. The reason health officials are concerned is because the virus is new, which makes it hard to predict how it will continue to affect people. Researchers and doctors are learning more about it every day, including exactly how it spreads and who is most at risk.

Short History:

In December 2019, a cluster of acute respiratory illness, now known as novel coronavirus–infected pneumonia (NCIP), occurred in Wuhan, Hubei Province, China. The disease has rapidly spread from Wuhan to other areas. As of January 31, 2020, a total of 9692 NCIP cases in China have been confirmed. Internationally, cases have been reported in 24 countries and 5 continents. On January 3, 2020, the 2019 novel coronavirus (2019-nCoV) was identified in samples of bron-choalveolar lavage fluid from a patient in Wuhan and was confirmed as the cause of the NCIP. Full-genome sequencing and phytogenic analysis indicated that 2019-nCoV is a distinct clade from the beta coronaviruses associated with human severe acute respiratory syndrome (SARS) and Middle East respiratory syndrome (MERS). The 2019-nCoV has features typical of the corona virus family and was classified in the beta coronavirus 2b lineage. The 2019-nCoV has close similarity to bat coronaviruses, and it has been postulated that bats are the primary source. While the origin of the 2019-nCoV is still being investigated, current evidence suggests spread to humans occurred via transmission from wild animals illegally sold in the Huanan Seafood Wholesale Market.

Huang et al first reported cases of NCIP in which most patients had a history of exposure to Huanan Seafood Wholesale Market. Patients' clinical manifestations included fever, non-productive cough, dyspnea, myalgia, fatigue, normal or decreased leukocyte counts, and radiographic evidence of pneumonia. Organ dysfunction (eg, shock, acute respiratory distress syndrome [ARDS], acute cardiac injury, and acute kidney injury) and death can occur in severe cases. Subsequently, Chen et al reported findings from 99 cases of NCIP from the same hospital and the results suggested that the 2019-nCoV infection clustered with in groups of humans in close contact, was more likely to affect older men with comorbidities, and could result in ARDS. However, the difference in clinical characteristics between severe and non-severe cases was not reported. Case reports confirmed human-to-human transmission of NCIP. At present, there are no effective therapies or vaccines for NCIP. The objective of this case series was to describe the clinical characteristics of 138 hospitalized patients with NCIP and to compare severe cases who received intensive care unit (ICU)c are with non-severe cases who did not receive ICU care.

Symptoms of a Coronavirus:

COVID-19 symptoms range from mild to severe. It takes 2-14 days after exposure for symptoms to develop. Symptoms may include:

Fever Difficulty in Breathing Coughing Tightness of Chest
Running Nose Head Ache Feeling of being Unwell Pneumonia
Kidney Failure.

Those with weakened immune systems may develop more serious symptoms, like pneumonia or bronchitis. You may never develop symptoms after being exposed to COVID-19. So far, most conrmed cases are in adults, but some children have been infected. There is no evidence that children are at greater risk for getting the virus.

Risk Persons:

According to the CDC (https://www.cdc.gov/coronavirus/2019-ncov/specific-groups/high-risk-complications.html), children do not seem to be at higher risk for getting COVID-19. However, some people are, including

- Older adults;

- People who have serious chronic medical conditions like: Heart disease Diabetes Lung disease Suppressed immune systems.

Causes a Coronavirus Infection:

Humans rst get a coronavirus from contact with animals. Then, it can spread from human to human. Health ocials do not know what animal caused COVID-19.

The COVID-19 virus can be spread through contact with certain bodily uids, such as droplets in a cough. It might also be caused by touching something an infected person has touched and then touching your hand to your mouth, nose, or eyes.

Mode of transmission:

Human Coronavirus (2019-nCoV) most commonly spread from an infected person to other through:

- The air by coughing and sneezing.
- Close personal contact, such as touching or shaking hand.
- Touching an object or surface with the virus on it, then touching your mouth, nose, or eyes before washing your hands.

Diagnosed of Coronavirus:

If you believe you have COVID-19, you should contact your family doctor immediately. Before going to the doctor's oce, call with your concerns. This will allow the oce to collect information and oer you guidance on next steps. To diagnose you, your doctor may run tests to rule out other common infections. In some cases, your doctor may suggest you self-isolate to prevent the spread of infection, etc.

Coronavirus be prevented or avoided:

Try to avoid people who are sick or meeting in large groups. Stay home if you are sick.

Cover your cough with a tissue or cough into your upper sleeve or elbow. Do not cough into your hands.

Wash your hands often with soap and water for at least 20 seconds, especially after going to the bathroom, before eating, and after blowing your nose, coughing, or sneezing. If soap and water are not readily available, use an alcohol-based hand sanitizer with at least 60% alcohol. Always wash hands with soap and water if hands are visibly dirty. Avoid touching your mouth, nose, or eyes. The Centers for Disease Control has issued travel advisories for several aected countries. If you are traveling to an area where the COVID-19 is present, speak with your doctor.

Reduce risk of Coronavirus infection (2019-nCoV):

- Clean hand with soap and water or alcohol based hand rub.
- Cover nose and mouth when coughing & sneezing with tissue or flexed elbow.
- Avoid close contact with anyone with cold or flu like symptoms.
- Avoid frozen meet.
- Isolation of symptomatic patients for atleast 14 days.

DO's and DON'T's:

DO's

- Cover your nose and mouth with disposable tissue or handkerchief while coughing or sneezing.
- Frequently wash your hands with soap and water.
- Avoid crowded places.
- Person suffering from Influenza like illness must be confined at home.
- Stay more than one arm's length distance from persons sick with flu.
- Take adequate sleep and rest.
- Drink plenty of water/liquids and eat nutritious food.
- Person suspected with Influenza like illness must consult doctor.

DON'T's

- Touching eyes, nose or mouth with unwashed hands.
- Hugging, kissing and shaking hands while greeting.
- Spitting in public places.
- Taking medicines without consulting doctor.
- Excessive physical exercise.
- Disposal of used napkin or tissue paper in open areas.
- Touching surfaces usually used by public (Railing, door gates, etc).
- Smoking in public places.
- Unnecessary testing.

Coronavirus treatment:

There is currently no vaccine or treatment for COVID-19. Symptoms of a coronavirus usually go away on their own. If symptoms feel worse than a common cold, contact your doctor. He or she may prescribe pain or fever medication.

As with a cold or the u, drink uids and get plenty of rest. If you are having trouble breathing, seek immediate medical care. When possible, avoid contact with others when you are sick. If you have COVID-19, wear a facemask to prevent spreading the virus to others. The CDC does not recommend wearing a mask if you do not have COVID-19.

About a Corona vaccine:

Scientists are working on a vaccine, but don't expect it anytime soon. The NIH is trying to develop one but says it will take at least a few months before clinical trials start and more than a year until a vaccine could become available. Separately, scientists in Texas, New York and China also are trying to create a vaccine, said Dr. Peter Hotez, a vaccine scientist at Baylor College of Medicine in Houston. But the challenge is daunting, Hotez said. "The lesson we've learned is coronavirus infections are serious and one of the newest and biggest global health threats." (CNN's Elizabeth Cohen, Dr. Sanjay Gupta and Jacqueline Howard contributed to this report.)

Protect your family:

There is currently no vaccine to prevent COVID-19, but there are a few things you can do to keep your family healthy:

- Wash your hands often with soap and water for at least 20 seconds. If soap and water are not available, use hand sanitizer Look for one that is 60% or higher alcohol-based.
- Keep your kids away from others who are sick or keep them home if they are ill.
- Teach kids to cough and sneeze into a tissue (make sure to throw it away after each use!) or to cough and sneeze into their arm or elbow, not their hands.
- Clean and disinfect your home as usual using regular household cleaning sprays or wipes.
- Avoid touching your face; teach your children to do the same.
- Avoid travel to highly infected areas.

To be prepare for the possibility of school or childcare closings:

If there is a COVID-19 outbreak in your community, local public health officials may decide to temporarily close schools and childcare centers to help slow the spread of the virus. Schools also may decide to dismiss students early if there are too many students or staff absent. Many schools use email to update families. Be sure your child's school knows how to get in touch with you.

Working parents can be prepared by having alternative childcare plans or talking with their employers about work from home options during school closings. If your child attends a college or university, encourage them to learn about the school's plan for a COVID-19 outbreak.

If your children need to stay at home due to the outbreak, try to keep their days as routine and scheduled as possible. Here are a few tips that can help:

- Read books with your child. It's not only fun, but reading together strengthens your bond with your child AND helps their development.
- Make time for active play. Bring out the blocks, balls, jump ropes and buckets and let the creativity go. Play games that kids of all ages can play, like tag or duck duck goose. Let your kids make up new games. Encourage older kids to make up a workout or dance to keep them moving.
- Keep an eye on media time. Whenever possible, play video games or go online with your child to keep that time structured and limited. If kids are missing their school friends or other family, try video chats to stay in touch.

Talking to children about COVID-19:

There's a lot of news coverage about the outbreak of COVID-19 and it can be overwhelming for parents and frightening to kids. The American Academy of Pediatrics encourages parents and others who work closely with children to filter information and talk about it in a way that their child can understand. These tips can help:

- Simple reassurance. Remind children that researchers and doctors are learning as much as they can, as quickly as they can, about the virus and are taking steps to keep everyone safe.

- Give them control. It's also a great time to remind your children of what they can do to help – washing their hands often, coughing into a tissue or their sleeves, and getting enough sleep.
- Watch for signs of anxiety. Children may not have the words to express their worry, but you may see signs of it. They may get cranky, be more clingy, have trouble sleeping, or seem distracted. Keep the reassurance going and try to stick to your normal routines.
- Monitor their media. Keep young children away from frightening images they may see on TV, social media, computers, etc. For older children, talk together about what they are hearing on the news and correct any misinformation or rumors you may hear.
- Be a good role model. COVID-19 doesn't discriminate and neither should we. While COVID-19 started in Wuhan, China, it doesn't mean that having Asian ancestry – or any other ancestry – makes someone more susceptible to the virus or more contagious. Stigma and discrimination hurt everyone by creating fear or anger towards others. When you show empathy and support to those who are ill, your children will too.

****

Chapter-3

COVID-19

Coronaviruses are a group of related viruses that cause diseases in mammals and birds. In humans, coronaviruses cause respiratory tract infections that can be mild, such as some cases of the common cold (among other possible causes, predominantly rhinoviruses), and others that can be lethal, such as SARS, MERS, and COVID-19. Symptoms in other species vary: in chickens, they cause an upper respiratory tract disease, while in cows and pigs they cause diarrhea. There are yet to be vaccines or antiviral drugs to prevent or treat human coronavirus infections.

Coronaviruses constitute the subfamily Orthocoronavirinae, in the family Coronaviridae, order Nidovirales, and realm Riboviria. They are enveloped viruses with a positive-sense single-stranded RNA genome and a nucleocapsid of helical symmetry. The genome size of coronaviruses ranges from approximately 27 to 34 kilobases, the largest among known RNA viruses. The name coronavirus is derived from the Latin corona, meaning "crown" or "halo", which refers to the characteristic appearance reminiscent of a crown or a solar corona around the virions (virus particles) when viewed under two-dimensional transmission electron microscopy, due to the surface being covered in club-shaped protein spikes.

Virus classification

(unranked):	Virus
Realm:	Riboviria
Phylum:	incertae sedis
Order:	Nidovirales
Family:	Coronaviridae
Subfamily:	Orthocoronavirinae

The World Health Organization announced in February 2020 that COVID-19 is the official name of the disease. World Health Organization chief Tedros Adhanom Ghebreyesus explained that CO stands for corona, VI for virus and D for disease, while 19 is for the year that the outbreak was first identified; 30 December 2019. The name had been chosen to avoid references to a specific geographical location (China), animal species, or group of people, in line with international recommendations for naming aimed at preventing stigmatisation.

While the disease is named COVID-19, the virus that causes it is named severe acute respiratory syndrome coronavirus 2 or SARS-CoV-2. The virus was initially referred to as the 2019 novel coronavirus or 2019-nCoV. The WHO additionally uses "the COVID-19 virus" and "the virus responsible for COVID-19" in public communications.

Coronaviruses were named in 1968 for their appearance in electron micrographs which was reminiscent of the solar corona, corona meaning crown in Latin.

Various Name of this virus:

- Coronavirus disease 2019.
- COVID-19.
- 2019-nCoV acute respiratory disease.
- Novel coronavirus pneumonia.
- "Coronavirus" or other names for SARS-CoV-2.

Discovery:

Human coronaviruses were first discovered in the late 1960s. The earliest ones discovered were an infectious bronchitis virus in chickens and two in human patients with the common cold (later named human coronavirus 229E and human coronavirus OC43). Other members of this family have since been identified, including SARS-CoV in 2003, HCoV NL63 in 2004, HKU1 in 2005, MERS-CoV in 2012, and SARS-CoV-2 (formerly known as 2019-nCoV) in 2019. Most of these have involved serious respiratory tract infections.

Etymology:

The name "coronavirus" is derived from Latin corona, meaning "crown" or "wreath", itself a borrowing from Greek "garland, wreath". The name refers to the characteristic appearance of virions (the infective form of the virus) by electron microscopy, which have a fringe of large, bulbous surface projections creating an image reminiscent of a crown or of a solar corona. This morphology is created by the viral spike peplomers, which are proteins on the surface of the virus.

Morphology:

Coronaviruses are large pleomorphic spherical particles with bulbous surface projections. The diameter of the virus particles is around 120 nm. The envelope of the virus in electron micrographs appears as a distinct pair of electron dense shells. The viral envelope consists of a lipid bilayer where the membrane (M), envelope (E) and spike (S) structural proteins are anchored. A subset of coronaviruses (specifically the members of betacoronavirus subgroup A) also have a shorter spike-like surface protein called hemagglutinin esterase (HE).

Inside the envelope, there is the nucleocapsid, which is formed from multiple copies of the nucleocapsid (N) protein, which are bound to the positive-sense single-stranded RNA genome in a continuous beads-on-a-string type conformation. The lipid bilayer envelope, membrane proteins, and nucleocapsid protect the virus when it is outside the host cell.

Genome:

Coronaviruses contain a positive-sense, single-stranded RNA genome. The genome size for coronaviruses ranges from approximately 27 to 34 kilobases. The genome size is one of the largest among RNA viruses. The genome has a 5′ methylated cap and a 3′ polyadenylated tail. The genome organization for a coronavirus is 5′-leader-UTR-replicase/transcriptase-spike (S)-envelope (E)-membrane (M)-nucleocapsid (N)-3′UTR-poly (A) tail. The open reading frames 1a and 1b, which occupy the first two-thirds of the genome, encode the replicase/transcriptase polyprotein. The replicase/transcriptase polyprotein self cleaves to form the nonstructural proteins (nsps).

The later reading frames encode the four major structural proteins: spike, envelope, membrane, and nucleocapsid. Interspersed between these reading frames are the reading frames for the accessory proteins. The number of accessory proteins and their function is unique depending on the specific coronavirus.

The life cycle of a coronavirus

- **Entry**

Infection begins when the viral spike (S) glycoprotein attaches to its complementary host cell receptor. After attachment, a protease of the host cell cleaves and activates the receptor-attached spike protein. Depending on the host cell protease available, cleavage and activation allows the virus to enter the host cell by endocytosis or direct fusion of the viral envelop with the host membrane.

On entry into the host cell, the virus particle is uncoated, and its genome enters the cell cytoplasm. The coronavirus RNA genome has a 5′ methylated cap and a 3′ polyadenylated tail, which allows the RNA to attach to the host cell's ribosome for translation. The host ribosome translates the initial overlapping open reading frame of the virus genome and forms a long polyprotein. The polyprotein has its own proteases which cleave the polyprotein into multiple nonstructural proteins.

- **Replication**

A number of the nonstructural proteins coalesce to form a multi-protein replicase-transcriptase complex (RTC). The main replicase-transcriptase protein is the RNA-dependent RNA polymerase (RdRp). It is directly involved in the replication and transcription of RNA from an RNA strand. The other nonstructural proteins in the complex assist in the replication and transcription process. The exoribonuclease non-structural protein, for instance, provides extra fidelity to replication by providing a proofreading function which the RNA-dependent RNA polymerase lacks.

One of the main functions of the complex is to replicate the viral genome. RdRp directly mediates the synthesis of negative-sense genomic RNA from the positive-sense genomic RNA. This is followed by the replication of positive-sense genomic RNA from the negative-sense genomic RNA. The other important function of the complex is to transcribe the viral genome. RdRp directly mediates the synthesis of negative-sense subgenomic RNA molecules from the positive-sense genomic RNA. This is followed by the transcription of these negative-sense subgenomic RNA molecules to their corresponding positive-sense mRNAs.

- **Release**

The replicated positive-sense genomic RNA becomes the genome of the progeny viruses. The mRNAs are gene transcripts of the last third of the virus genome after the initial overlapping reading frame. These mRNAs are translated by the host's ribosomes into the structural proteins and a number of accessory proteins. RNA translation occurs inside the endoplasmic reticulum. The viral structural proteins S, E, and M move along the secretory pathway into the Golgi intermediate compartment. There, the M proteins direct most protein-protein interactions required for assembly of viruses following its binding to the nucleocapsid. Progeny viruses are then released from the host cell by exocytosis through secretory vesicles.

Transmission:

Human to human transmission of coronaviruses is primarily thought to occur among close contacts via respiratory droplets generated by sneezing and coughing. The interaction of the coronavirus spike protein with its complement host cell receptor is central in determining the tissue tropism, infectivity, and species range of the virus. The SARS coronavirus, for example, infects human cells by attaching to the angiotensin-converting enzyme 2 (ACE2) receptor.

Taxonomy:

The scientific name for coronavirus is Orthocoronavirinae or Coronavirinae. Coronavirus belongs to the family of Coronaviridae.

- **Genus:** Alphacoronavirus

Species: Human coronavirus 229E, Human coronavirus NL63, Miniopterus bat coronavirus 1, Miniopterus bat coronavirus HKU8, Porcine epidemic diarrhea virus, Rhinolophus bat coronavirus HKU2, Scotophilus bat coronavirus 512

- **Genus** Betacoronavirus; type species: Murine coronavirus

Species: Betacoronavirus 1, Human coronavirus HKU1, Murine coronavirus, Pipistrellus bat coronavirus HKU5, Rousettus bat coronavirus HKU9, Severe acute respiratory syndrome-related coronavirus, Severe acute respiratory syndrome coronavirus 2, Tylonycteris bat coronavirus HKU4, Middle East respiratory syndrome-related coronavirus, Human coronavirus OC43, Hedgehog coronavirus 1 (EriCoV)

- **Genus** Gammacoronavirus; type species: Infectious bronchitis virus

Species: Beluga whale coronavirus SW1, Infectious bronchitis virus

- **Genus** Deltacoronavirus; type species: Bulbul coronavirus HKU11

Species: Bulbul coronavirus HKU11, Porcine coronavirus HKU15

Evolution:

The most recent common ancestor (MRCA) of all coronaviruses has been estimated to have existed as recently as 8000 BCE, though some models place the MRCA as far back as 55 million years or more, implying long term coevolution with bats. The MRCAs of the alphacoronavirus line has been placed at about 2400 BCE, the betacoronavirus line at 3300 BCE, the gammacoronavirus line at 2800 BCE, and the deltacoronavirus line at about 3000 BCE. It appears that bats and birds, as warm-blooded flying vertebrates, are ideal hosts for the coronavirus gene source (with bats for alphacoronavirus and betacoronavirus, and birds for gammacoronavirus and deltacoronavirus) to fuel coronavirus evolution and dissemination.

Bovine coronavirus and canine respiratory coronaviruses diverged from a common ancestor recently (-1950). Bovine coronavirus and human coronavirus OC43 diverged around the 1890s. Bovine coronavirus diverged from the equine coronavirus species at the end of the 18th century.

The MRCA of human coronavirus OC43 has been dated to the 1950s. MERS-CoV, although related to several bat coronavirus species, appears to have diverged from these several centuries ago. The human coronavirus NL63 and a bat coronavirus shared an MRCA 563–822 years ago.

The most closely related bat coronavirus and SARS-CoV diverged in 1986. A path of evolution of the SARS virus and keen relationship with bats have been proposed. The authors suggest that the coronaviruses have been coevolved with bats for a long time and the ancestors of SARS-CoV first infected the species of the genus Hipposideridae, subsequently spread to species of the Rhinolophidae and then to civets, and finally to humans. Alpaca coronavirus and human coronavirus 229E diverged before 1960.

Human coronaviruses:

Coronaviruses vary significantly in risk factor. Some can kill more than 30% of those infected (such as MERS-CoV), and some are relatively harmless, such as the common cold. Coronaviruses cause colds with major symptoms, such as fever, and sore throat from swollen adenoids, occurring primarily in the winter and early spring seasons. Coronaviruses can cause pneumonia (either direct viral pneumonia or a secondary bacterial pneumonia) and bronchitis (either direct viral bronchitis or a secondary bacterial bronchitis). The much publicized human coronavirus discovered in 2003, SARS-CoV, which causes severe acute respiratory syndrome (SARS), has a unique pathogenesis because it causes both upper and lower respiratory tract infections.

Seven strains of human coronaviruses are known, of which four produce the generally mild symptoms of the common cold:

- Human coronavirus OC43 (HCoV-OC43)
- Human coronavirus HKU1
- Human coronavirus NL63 (HCoV-NL63, New Haven coronavirus)
- Human coronavirus 229E (HCoV-229E)

and three, symptoms that are potentially severe:

- Middle East respiratory syndrome-related coronavirus (MERS-CoV), previously known as novel coronavirus 2012 and HCoV-EMC
- Severe acute respiratory syndrome coronavirus (SARS-CoV or "SARS-classic")
- Severe acute respiratory syndrome coronavirus 2 (SARS-CoV-2), previously known as 2019-nCoV or "novel coronavirus 2019"

The coronaviruses HCoV-229E, -NL63, -OC43, and -HKU1 continually circulate in the human population and cause respiratory infections in adults and children world-wide.

Outbreaks of coronavirus-related diseases:

Outbreaks of coronavirus types of relatively high mortality are as follows:

Outbreak	**Virus type**	**Deaths**
2003 severe acute respiratory syndrome outbreak	SARS-CoV	774
2012 Middle East respiratory syndrome coronavirus outbreak	MERS-CoV	Over 400
2015 Middle East respiratory syndrome outbreak in South Korea	MERS-CoV	36
2018 Middle East respiratory syndrome outbreak	MERS-CoV	41
2019–2020 coronavirus pandemic	SARS-CoV-2	At least 18,883 (as per date of 23rd March of 2020 but its increasing.)

- **Severe acute respiratory syndrome (SARS)**

In 2003, following the outbreak of severe acute respiratory syndrome (SARS) which had begun the prior year in Asia, and secondary cases elsewhere in the world, the World Health Organization (WHO) issued a press release stating that a novel coronavirus identified by a number of laboratories was the causative agent for SARS. The virus was officially named the SARS coronavirus (SARS-CoV). More than 8,000 people were infected, about ten percent of whom died.

- **Middle East respiratory syndrome (MERS)**

In September 2012, a new type of coronavirus was identified, initially called Novel Coronavirus 2012, and now officially named Middle East respiratory syndrome coronavirus (MERS-CoV). The World Health Organization issued a global alert soon after. The WHO update on 28 September 2012 said the virus did not seem to pass easily from person to person. However, on 12 May 2013, a case of human-to-human transmission in France was confirmed by the French Ministry of Social Affairs and Health. In addition, cases of human-to-human transmission were reported by the Ministry of Health in Tunisia. Two confirmed cases involved people who seemed to have caught the disease from their late father, who became ill after a visit to Qatar and Saudi Arabia. Despite this, it appears the virus had trouble spreading from human to human, as most individuals who are infected do not transmit the virus. By 30 October 2013, there were 124 cases and 52 deaths in Saudi Arabia.

After the Dutch Erasmus Medical Centre sequenced the virus, the virus was given a new name, Human Coronavirus–Erasmus Medical Centre (HCoV-EMC). The final name for the virus is Middle East respiratory syndrome coronavirus (MERS-CoV). In May 2014, the only two United States cases of MERS-CoV infection were recorded, both occurring in healthcare workers who worked in Saudi Arabia and then travelled to the U.S. One was treated in Indiana and one in Florida. Both were hospitalized temporarily and then discharged.

In May 2015, an outbreak of MERS-CoV occurred in the Republic of Korea, when a man who had traveled to the Middle East, visited 4 hospitals in the Seoul area to treat his illness. This caused one of the largest outbreaks of MERS-CoV outside the Middle East. As of December 2019, 2,468 cases of MERS-CoV infection had been confirmed by laboratory tests, 851 of which were fatal, a mortality rate of approximately 34.5%.

- **Coronavirus disease 2019 (COVID-19):**

In December 2019, a pneumonia outbreak was reported in Wuhan, China. On 31 December 2019, the outbreak was traced to a novel strain of coronavirus, which was given the interim name 2019-nCoV by the World Health Organization (WHO), later renamed SARS-CoV-2 by the International Committee on Taxonomy of Viruses. Some researchers have suggested that the Huanan Seafood Wholesale Market may not be the original source of viral transmission to humans.

As of 25 March 2020, there have been at least 18,883 confirmed deaths and more than 421,792 confirmed cases in the coronavirus pneumonia pandemic. The Wuhan strain has been identified as a new strain of Betacoronavirus from group 2B with approximately 70% genetic similarity to the SARS-CoV. The virus has a 96% similarity to a bat coronavirus, so it is widely suspected to originate from bats as well. The pandemic has resulted in travel restrictions and nationwide lockdowns in several countries.

Other animals:

Coronaviruses have been recognized as causing pathological conditions in veterinary medicine since the early 1970s. Except for avian infectious bronchitis, the major related diseases have mainly an intestinal location.

Diseases caused

Coronaviruses primarily infect the upper respiratory and gastrointestinal tract of mammals and birds. They also cause a range of diseases in farm animals and domesticated pets, some of which can be serious and are a threat to the farming industry. In chickens, the infectious bronchitis virus (IBV), a coronavirus, targets not only the respiratory tract but also the urogenital tract. The virus can spread to different organs throughout the chicken. Economically significant coronaviruses of farm animals include porcine coronavirus (transmissible gastroenteritis coronavirus, TGE) and bovine coronavirus, which both result in diarrhea in young animals. Feline coronavirus: two forms, feline enteric coronavirus is a pathogen of minor clinical significance, but spontaneous mutation of this virus can result in feline infectious peritonitis (FIP), a disease associated with high mortality. Similarly, there are two types of coronavirus that infect ferrets: Ferret enteric coronavirus causes a gastrointestinal syndrome known as epizootic catarrhal enteritis (ECE), and a more lethal systemic version of the virus (like FIP in cats) known as ferret systemic coronavirus (FSC). There are two types of canine coronavirus (CCoV), one that causes mild gastrointestinal disease and one that has been found to cause respiratory disease. Mouse hepatitis virus (MHV) is a coronavirus that causes an epidemic murine illness with high mortality, especially among colonies of laboratory mice. Sialodacryoadenitis virus (SDAV) is highly infectious coronavirus of laboratory rats, which can be transmitted between individuals by direct contact and indirectly by aerosol. Acute infections have high morbidity and tropism for the salivary, lachrymal and harderian glands.

A HKU2-related bat coronavirus called swine acute diarrhea syndrome coronavirus (SADS-CoV) causes diarrhea in pigs.

Prior to the discovery of SARS-CoV, MHV had been the best-studied coronavirus both in vivo and in vitro as well as at the molecular level. Some strains of MHV cause a progressive demyelinating encephalitis in mice which has been used as a murine model for multiple sclerosis. Significant research efforts have been focused on elucidating the viral pathogenesis of these animal coronaviruses, especially by virologists interested in veterinary and zoonotic diseases.

In domestic animals

- Infectious bronchitis virus (IBV) causes avian infectious bronchitis.
- Porcine coronavirus (transmissible gastroenteritis coronavirus of pigs, TGEV).
- Bovine coronavirus (BCV), responsible for severe profuse enteritis in of young calves.
- Feline coronavirus (FCoV) causes mild enteritis in cats as well as severe Feline infectious peritonitis (other variants of the same virus).
- the two types of canine coronavirus (CCoV) (one causing enteritis, the other found in respiratory diseases).
- Turkey coronavirus (TCV) causes enteritis in turkeys.
- Ferret enteric coronavirus causes epizootic catarrhal enteritis in ferrets.
- Ferret systemic coronavirus causes FIP-like systemic syndrome in ferrets.
- Pantropic canine coronavirus.
- Rabbit enteric coronavirus causes acute gastrointestinal disease and diarrhea in young European rabbits. Mortality rates are high.
- Porcine epidemic diarrhea virus (PED or PEDV), has emerged around the world.

Genomic cis-acting elements:

In common with the genomes of all other RNA viruses, coronavirus genomes contain cis-acting RNA elements that ensure the specific replication of viral RNA by a virally encoded RNA-dependent RNA polymerase. The embedded cis-acting elements devoted to coronavirus replication constitute a small fraction of the total genome, but this is presumed to be a reflection of the fact that coronaviruses have the largest genomes of all RNA viruses. The boundaries of cis-acting elements essential to replication are fairly well-defined, and the RNA secondary structures of these regions are understood. However, how these cis-acting structures and sequences interact with the viral replicase and host cell components to allow RNA synthesis is not well understood.

Genome packaging

The assembly of infectious coronavirus particles requires the selection of viral genomic RNA from a cellular pool that contains an abundant excess of non-viral and viral RNAs. Among the seven to ten specific viral mRNAs synthesized in virus-infected cells, only the full-length genomic RNA is packaged efficiently into coronavirus particles. Studies have revealed cis-acting elements and trans-acting viral factors involved in the coronavirus genome encapsidation and packaging. Understanding the molecular mechanisms of genome selection and packaging is critical for developing antiviral strategies and viral expression vectors based on the coronavirus genome.

****

Chapter- 4

Symptoms, Causes, Diagnosis, Prevention, Management, and Research on COVID-19

Introduction:

Coronavirus disease 2019 (COVID-19) is an infectious disease caused by severe acute respiratory syndrome coronavirus 2 (SARS-CoV-2). The disease was first identified in 2019 in Wuhan, the capital of Hubei, China, and has since spread globally, resulting in the 2019–20 coronavirus pandemic. Common symptoms include fever, cough, and shortness of breath. Muscle pain, sputum production, diarrhea, and sore throat are less common. While the majority of cases result in mild symptoms, some progress to pneumonia and multi-organ failure. As of 23 March 2020, the rate of deaths per number of diagnosed cases is 4.4 percent; however, it ranges from 0.2 percent to 15 percent, according to age group and other health problems.

The virus is typically spread during close contact and via respiratory droplets produced when people cough or sneeze. Respiratory droplets may be produced during breathing but it is not considered airborne. It may also spread when one touches a contaminated surface and then their face. It is most contagious when people are symptomatic, although spread may be possible before symptoms appear. The virus can live on surfaces up to 72 hours. Time from exposure to onset of symptoms is generally between two and fourteen days, with an average of five days. The standard method of diagnosis is by reverse transcription polymerase chain reaction (rRT-PCR) from a nasopharyngeal swab. The infection can also be diagnosed from a combination of symptoms, risk factors and a chest CT scan showing features of pneumonia.

Recommended measures to prevent infection include frequent hand washing, social distancing (maintaining physical distance from others), and keeping hands away from the face. The use of masks is recommended for those who suspect they have the virus and their caregivers, but not for the general public, although simple cloth masks may be used by those who desire them. There is no vaccine or specific antiviral treatment for COVID-19. Management involves treatment of symptoms, supportive care, isolation, and experimental measures.

The World Health Organization (WHO) declared the 2019–20 coronavirus outbreak a Public Health Emergency of International Concern (PHEIC) on 30 January 2020 and a pandemic on 11 March 2020. Evidence of local transmission of the disease has been found in many countries across all six WHO regions.

Signs and symptoms:

Although those infected with the virus may be asymptomatic, many develop flu-like symptoms, including fever, cough, and shortness of breath. Emergency symptoms include difficulty breathing, persistent chest pain or pressure, confusion, difficulty waking, Fever, Dry cough, Fatigue, Sputum production, Loss of smell, Shortness of breath, Muscle or joint pain, Sore throat, Headache, Chills, Nausea or vomiting, Nasal congestion, Diarrhea, Haemoptysis, Conjunctival congestion and bluish face or lips; immediate medical attention is advised if these symptoms are present. Less commonly, upper respiratory symptoms, such as sneezing, runny nose, or sore throat may be seen. Symptoms such as nausea, vomiting, and diarrhea have been observed in varying percentages among people in several studies, with percentages varying from 3% to 31% of cases depending on the study. Some initial cases in China presented only chest tightness and palpitations. In some, the disease may progress to pneumonia, multi-organ failure, and death.

As it is common with infections, there is a delay from when a person is infected with the virus to when they develop symptoms, known as the incubation period. The incubation period for COVID-19 is typically five to six days but may range from two to 14 days. 97.5% of people who develop symptoms will do so within 11.5 days of infection.

Cause:

The disease is caused by the virus severe acute respiratory syndrome coronavirus 2 (SARS-CoV-2), previously referred to as the 2019 novel coronavirus (2019-nCoV). It is primarily spread between people via respiratory droplets from coughs and sneezes. A study investigating the rate of decay of the virus found no viable viruses after 4 h on copper, 24 h on cardboard, 72 h on stainless steel, and 72 h on plastic. However, detection rates did not reach 100% and varied between surface type (limit of detection was 3.33×100.5 TCID50 per liter of air for aerosols, 100.5 TCID50 per milliliter of medium for plastic, steel, and cardboard, and 101.5 TCID50 per milliliter of medium for copper). Estimation of the rate of decay with a Bayesian regression model suggests that viruses may remain viable up to 18 h on copper, 55 h on cardboard, 90 h on stainless steel, and over 100 h on plastic. The virus remained viable in aerosols throughout the time of the experiment (3 h). The virus has also been found in faeces, and transmission through faeces is being researched.

The lungs are the organs most affected by COVID-19 because the virus accesses host cells via the enzyme ACE2, which is most abundant in the type II alveolar cells of the lungs. The virus uses a special surface glycoprotein called a "spike" (peplomer) to connect to ACE2 and enter the host cell. The density of ACE2 in each tissue correlates with the severity of the disease in that tissue and some have suggested that decreasing ACE2 activity might be protective, though another view is that increasing ACE2 using angiotensin II receptor blocker medications could be protective and that these hypotheses need to be tested. As the alveolar disease progresses, respiratory failure might develop and death may follow.

The virus also affects gastrointestinal organs as ACE2 is abundantly expressed in the glandular cells of gastric, duodenal and rectal epithelium as well as endothelial cells and enterocytes of the small intestine. The virus has been found in the faeces of as many as 53% of hospitalised people and more anal swab positives have been found than oral swab positives in the later stages of infection. The virus was found in faeces from 1 to 12 days and 17% of patients continued to present the virus in faeces after no longer presenting them in respiratory samples, indicating that the viral gastrointestinal infection and the potential fecal-oral transmission can last even after viral clearance in the respiratory tract. Reoccurrence of the virus has also been detected through anal swabs suggesting a shift from more oral positive during the early stages of the disease to more anal positive during later periods.

The virus is thought to be natural and have an animal origin, through spillover infection. The origin is unknown but by December 2019 the spread of infection was almost entirely driven by human-to-human transmission. The earliest known infection occurred on 17 November 2019 in Wuhan, China.

Diagnosis:

The WHO has published several testing protocols for the disease. The standard method of testing is real-time reverse transcription polymerase chain reaction (rRT-PCR). The test can be done on respiratory samples obtained by various methods, including a nasopharyngeal swab or sputum sample. Results are generally available within a few hours to two days. Blood tests can be used, but these require two blood samples taken two weeks apart and the results have little immediate value. Chinese scientists were able to isolate a strain of the coronavirus and publish the genetic sequence so that laboratories across the world could independently develop polymerase chain reaction (PCR) tests to detect infection by the virus. As of 19 March 2020, there were no antibody tests though efforts to develop them are ongoing. The FDA approved the first point-of-care test on 21 March 2020 for use at the end of that month.

Diagnostic guidelines released by Zhongnan Hospital of Wuhan University suggested methods for detecting infections based upon clinical features and epidemiological risk. These involved identifying people who had at least two of the following symptoms in addition to a history of travel to Wuhan or contact with other infected people: fever, imaging features of pneumonia, normal or reduced white blood cell count, or reduced lymphocyte count.

One study in China found that CT scans showed ground-glass opacities in 56%, but 18% had no radiological findings. Bilateral and peripheral ground glass opacities are the most typical CT findings, though they are non-specific. Consolidation, linear opacities and reverse halo sign are other radiological findings. Initially, the lesions are confined to one lung, but as the disease progresses, indications manifest in both lungs in 88% of so-called "late patients" in the study group (the subset for whom time between onset of symptoms and chest CT was 6–12 days). Ground glass opacities are also a common feature in children's disease.

Prevention:

Preventive measures to reduce the chances of infection include staying at home, avoiding crowded places, washing hands with soap and warm water often and for at least 20 seconds, practicing good respiratory hygiene and avoiding touching the eyes, nose, or mouth with unwashed hands. The CDC recommends covering the mouth and nose with a tissue when coughing or sneezing, or using inside of the elbow if no tissue is available. They also recommend proper hand hygiene after any cough or sneeze. Social distancing strategies aim to reduce contact of infected persons with large groups by closing schools and workplaces, restricting travel, and canceling mass gatherings. Social distancing also includes that people stay at least 6 feet apart (about 1.80 meters). Because a vaccine against SARS-CoV-2 is not expected to become available until 2021 at the earliest, a key part of managing the COVID-19 pandemic is trying to decrease the epidemic peak, known as flattening the epidemic curve through various measures seeking to reduce the rate of new infections. Slowing the infection rate helps decrease the risk of health services being overwhelmed, allowing for better treatment of current cases, and delaying additional cases until therapeutics or a vaccine become available.

According to the WHO, the use of masks is recommended only if a person is coughing or sneezing or when one is taking care of someone with a suspected infection. Some countries also recommend healthy individuals to wear face masks, particularly China, Hong Kong and Thailand. In order to meet the need for masks, the WHO estimates that global production will need to increase by 40%. Hoarding and speculation have worsened the problem, with the price of masks increasing sixfold, N95 respirators tripled, and gowns doubled. Some health experts consider wearing non-medical-grade masks and other face coverings like scarves or bandanas a good way to prevent people from touching their mouths and noses, even if non-medical coverings would not protect against a direct sneeze or cough from an infected person.

Those diagnosed with COVID-19 or who believe they may be infected are advised by the CDC to stay home except to get medical care, call ahead before visiting a healthcare provider, wear a face mask when exposed to an individual or location of a suspected infection, cover coughs and sneezes with a tissue, regularly wash hands with soap and water and avoid sharing personal household items. The CDC also recommends that individuals wash hands often with soap and water for at least 20 seconds, especially after going to the toilet or when hands are visibly dirty, before eating and after blowing one's nose, coughing, or sneezing. It further recommends using an alcohol-based hand sanitizer with at least 60% alcohol, but only when soap and water are not readily available.

For areas where commercial hand sanitizers are not readily available, the WHO suggested two formulations for the local production. In both these formulations the antimicrobial activity of ethanol or isopropanol is enhanced by a low concentration of hydrogen peroxide while glycerol acts as a humectant.

Management:

People are managed with supportive care, which may include fluid, oxygen support, and supporting other affected vital organs. Steroids such as methylprednisolone are not recommended unless the disease is complicated by acute respiratory distress syndrome.

The CDC recommends that those who suspect they carry the virus wear a simple face mask. Extracorporeal membrane oxygenation (ECMO) has been used to address the issue of respiratory failure, but its benefits are still under consideration. While WHO does not oppose the use of non-steroidal anti-inflammatory drugs such as ibuprofen for symptoms, some recommend paracetamol (acetaminophen) for first-line use. There are four reported cases of children who developed severe symptoms after taking ibuprofen. While theoretical concerns have been raised about ACE inhibitors and Angiotensin receptor blockers, as of 19 March 2020 these are not sufficient to justify stopping these medications.

The WHO and Chinese National Health Commission have published recommendations for taking care of people who are hospitalised with COVID-19. Intensivists and pulmonologists in the US have compiled treatment recommendations from various agencies into a free resource, the IBCC.

Personal protective equipment

Precautions must be taken to minimize the risk of virus transmission, especially in healthcare settings when performing procedures that can generate aerosols, such as intubation or hand ventilation. Some public health officials have complained that the limited supply of protective gear should not be used to expand outpatient testing. New York City has departed from CDC guidelines and is recommending that testing be limited to hospitalized patients to conserve supplies. CDC outlines the specific guidelines for the use of personal protective equipment (PPE) during the pandemic. The recommended gear is as follows:

- Gown;
- mask or respirator,
- goggles or a face shield,

- ➢ medical gloves

The N95 respirators are approved for industrial settings but the FDA has authorized the masks for use under an Emergency Use Authorization (EUA). They are designed to protect from airborne particles like dust but effectiveness against a specific biological agent is not guaranteed for off-label uses. When masks are not available the CDC recommends using face shields, or as a last resort homemade masks.

Mechanical ventilation:

Most cases of COVID-19 are not severe enough to require mechanical ventilation (artificial assistance to support breathing), but a percentage of cases do. Some Canadian doctors recommend the use of invasive mechanical ventilation because this technique limits the spread of aerosolized transmission vectors. Severe cases are most common in older adults (those older than 60 years and especially those older than 80 years. Many developed countries do not have enough hospital beds per capita, which limits a health system's capacity to handle a sudden spike in the number of COVID-19 cases severe enough to require hospitalization. This limited capacity is a significant driver of the need to flatten the curve (to keep the speed at which new cases occur and thus the number of people sick at one point in time lower). One study in China found 5% were admitted to intensive care units, 2.3% needed mechanical support of ventilation, and 1.4% died. An Italian startup employed 3D printing technology to produce valves for life-saving coronavirus treatment due to a broken supply chain of original manufacturing.

Experimental treatment:

No medications are approved to treat the disease by the WHO although some are recommended by individual national medical authorities. Research into potential treatments started in January 2020, and several antiviral drugs are in clinical trials. Although new medications may take until 2021 to develop, several of the medications being tested are already approved for other uses, or are already in advanced testing. Antiviral medication may be tried in people with severe disease. The WHO recommended volunteers take part in trials of the effectiveness and safety of potential treatments.

Information technology:

In February 2020, China launched a mobile app to deal with the disease outbreak. Users are asked to enter their name and ID number. The app is able to detect 'close contact' using surveillance data and therefore a potential risk of infection. Every user can also check the status of three other users. If a potential risk is detected, the app not only recommends self-quarantine, it also alerts local health officials.

Big data analytics on cellphone data, facial recognition technology, mobile phone tracking and artificial intelligence are used to track infected people and people whom they contacted in South Korea, Taiwan and Singapore. In March 2020, the Israeli government enabled security agencies to track mobile phone data of people supposed to have coronavirus. The measure was taken to enforce quarantine and protect those who may come into contact with infected citizens. Also in March 2020, Deutsche Telekom shared private cellphone data with the German federal government agency, Robert Koch Institute, in order to research and prevent the spread of the virus. Russia deployed facial recognition technology to detect quarantine breakers. Italian regional health commissioner Giulio Gallera said that he has been informed by mobile phone operators that "40% of people are continuing to move around anyway". German government conducted a 48 hours weekend hackathon with more than 42.000 participants. Also the president of Estonia, Kersti Kaljulaid, made a global call for creative solutions against the spread of coronavirus.

Psychological support:

Infected individuals may experience distress from quarantine, travel restrictions, side effects of treatment, or fear of the infection itself. To address these concerns, the National Health Commission of China published a national guideline for psychological crisis intervention on 27 January 2020.

Prognosis:

The severity of COVID-19 varies. The disease may take a mild course with few or no symptoms, resembling other common upper respiratory diseases such as the common cold. Mild cases typically recover within two weeks, while those with severe or critical disease may take three to six weeks to recover. Among those who have died, the time from symptom onset to death has ranged from two to eight weeks.

Children of all ages are susceptible to the disease, but are likely to have milder symptoms and a much lower chance of severe disease than adults; in those younger than 50 years, the risk of death is less than 0.5%, while in those older than 70 it is more than 8%. Pregnant women are at particular risk for severe infection.

In some people, COVID-19 may affect the lungs causing pneumonia. In those most severely affected, COVID-19 may rapidly progress to acute respiratory distress syndrome (ARDS) causing respiratory failure, septic shock, or multi-organ failure. Complications associated with COVID-19 include sepsis, abnormal clotting, and damage to the heart, kidneys, and liver. Clotting abnormalities, specifically an increase in prothrombin time, have been described in 6% of those admitted to hospital with COVID-19, while abnormal kidney function is seen in 4% of this group. Liver injury as shown by blood markers of liver damage is frequently seen in severe cases. Some studies have found that the neutrophil to lymphocyte ratio (NLR) may be helpful in early screening for severe illness.

Many of those who die of COVID-19 have preexisting conditions, including hypertension, diabetes mellitus, and cardiovascular disease. The Istituto Superiore di Sanità reported that 68.9% of deaths from the disease in the country had at least one preexisting condition, with patients that had preexisting conditions having an average of 2.7 conditions. According to the same report, the median time between onset of symptoms and death was eight days, with half that time being spent hospitalized. However, patients transferred to an ICU had a median time of five days between hospitalization and death. In a study of early cases, the median time from exhibiting initial symptoms to death was 14 days, with a full range of six to 41 days. In a study by the National Health Commission (NHC) of China, men had a death rate of 2.8% while women had a death rate of 1.7%. Histopathological examinations of post-mortem lung samples show diffuse alveolar damage with cellular fibromyxoid exudates in both lungs. Viral cytopathic changes were observed in the pneumocytes. The lung picture resembled acute respiratory distress syndrome (ARDS). In 11.8% of the deaths reported by the National Health Commission of China, heart damage was noted by elevated levels of troponin or cardiac arrest.

Availability of medical resources and the socioeconomics of a region may also affect mortality. Estimates of the mortality from the condition vary because of those regional differences, but also because of methodological difficulties. The under-counting of mild cases can cause the mortality rate to be overestimated. However, the fact that deaths are the result of cases contracted in the past can mean the current mortality rate is underestimated.

It is unknown if past infection provides effective and long-term immunity in people who recover from the disease. Immunity is likely, based on the behaviour of other coronaviruses, but cases in which recovery from COVID-19 have been followed by positive tests for coronavirus at a later date have been reported. It is unclear if these cases are the result of reinfection, relapse, or testing error. Concerns have been raised about long-term sequelae of the disease. The Hong Kong Hospital Authority found a drop of 20% to 30% in lung capacity in some people who recovered from the disease, and lung scans suggested organ damage.

Epidemiology:

The case fatality rate (CFR) depends on the availability of healthcare, the typical age and health problems within the population, and the number of undiagnosed cases. Preliminary research has yielded case fatality rate numbers between 2% and 3%; in January 2020 the WHO suggested that the case fatality rate was approximately 3%, and 2% in February 2020 in Hubei. Other CFR numbers, which adjust for differences in time of confirmation, death or remission but are not peer reviewed, are respectively 7% and 33% for people in Wuhan 31 January. An unreviewed preprint of 55 deaths noted that early estimates of mortality may be too high as asymptomatic infections are missed. They estimated a mean infection fatality ratio (IFR, the mortality among infected) ranging from 0.8% to 0.9%. A peer-reviewed article published on 19 March estimated the overall symptomatic case fatality risk as 1.4% (IQR 0.9–2.1%). The outbreak in 2019–2020 has caused at least 421,792edit confirmed infections and 18,883edit deaths.

The epidemic spreads faster where people are close together and/or travel to other areas. Researchers found that travel restrictions can reduce the basic reproduction number from 2.35 to 1.05, allowing the epidemic to be manageable.

An observational study of nine people found no vertical transmission from mother to the newborn. Also, a descriptive study in Wuhan found no evidence of viral transmission through vaginal sex (from female to partner), but authors note that transmission during sex might occur through other routes.

Research:

- **Vaccine**

A COVID-19 vaccine is a hypothetical vaccine against coronavirus disease 2019 (COVID-19). Although no vaccine has completed clinical trials, there are multiple attempts in progress to develop such a vaccine. In late February 2020, the World Health Organization (WHO) said it did not expect a vaccine against SARS-CoV-2, the causative virus, to become available in less than 92 months. By early March 2020, some 30 vaccine candidates were in development.

Previous efforts:

Vaccines have been produced against several diseases caused by coronaviruses for animal use, including for infectious bronchitis virus in birds, canine coronavirus and feline coronavirus. Previous efforts to develop vaccines for viruses in the family Coronaviridae that affect humans have been aimed at severe acute respiratory syndrome coronavirus (SARS) and Middle East respiratory syndrome (MERS). Vaccines against SARS and MERS have been tested in non-human animal models. As of 2020, there is no cure or protective vaccine for SARS that has been shown to be both safe and effective in humans. According to research papers published in 2005 and 2006, the identification and development of novel vaccines and medicines to treat SARS was a priority for governments and public health agencies around the world.

There is also no proven vaccine against MERS. When MERS became prevalent, it was believed that existing SARS research may provide a useful template for developing vaccines and therapeutics against a MERS-CoV infection. As of March 2020, there was one (DNA based) MERS vaccine which completed phase I clinical trials in humans, and three others in progress, all of which are viral vectored vaccines, two adenoviral-vectored (ChAdOx1-MERS, BVRS-GamVac), and one MVA-vectored (MVA-MERS-S).

2020 efforts:

SARS-CoV-2 was identified on 1 December 2019 as the cause of what would later be named COVID-19. A major outbreak spread around the world in 2020, leading to considerable investment and research activity to develop a vaccine. Many organizations are using published genomes to develop possible vaccines against SARS-CoV-2. About 35 companies and academic institutions are involved, with three of them receiving support from the Coalition for Epidemic Preparedness Innovations (CEPI), including projects by the biotechnology companies Moderna, and Inovio Pharmaceuticals, and the University of Queensland. Some 300 early-stage studies are in progress worldwide, according to one report on 10 March 2020.

In early March 2020, CEPI announced a US$2 billion funding goal in a global partnership between public, private, philanthropic, and civil society organisations to accelerate development of COVID-19 vaccines, with commitments to date by the governments of Denmark, Finland, Germany, Norway, and the UK.

Clinical trials in progress

The US National Institute of Allergy and Infectious Disease (NIAID) collaborated with Moderna to develop an RNA vaccine matching a spike of the coronavirus surface. NIAID registered a Phase I safety clinical trial of the vaccine candidate, called mRNA-1273, to study 45 healthy adults in Seattle, with recruitment completed on 19 March 2020. An additional site at Emory University (Atlanta, Georgia) will start in April.

A Phase I safety trial of a recombinant adenovirus vaccine candidate manufactured by CanSino Biologics Inc. (Tianjin, China), called Ad5-nCoV, began recruiting 108 healthy adults in Wuhan, China in March, with the study to last through the end of 2020.

Limitations

It is possible vaccines in development will not be safe or effective. One study found that between 2006 to 2015, the success rate of obtaining approval from Phase I to successful Phase III trials was 11.5% for vaccines ("biologics").

Reports of preclinical research

In Australia, the University of Queensland announced that it is investigating the potential of a molecular clamp vaccine that would genetically modify viral proteins in order to stimulate an immune reaction.

Vaccine development efforts were announced at the Chinese Center for Disease Control and Prevention, and the University of Hong Kong. Around 29 January 2020, Janssen Pharmaceutical Companies, led by Hanneke Schuitemaker, announced that it had begun work on developing a vaccine. Janssen is codeveloping an oral vaccine with its biotechnology partner, Vaxart. On 18 March 2020, Emergent BioSolutions announced a manufacturing partnership with Vaxart to develop the vaccine. Inovio Pharmaceuticals said that it is developing a DNA-based vaccination in collaboration with a Chinese firm, announcing plans for human clinical trials in the summer of the Northern Hemisphere of 2020.

In early March, the Canadian government announced Can$275 million in funding for 96 research projects on medical countermeasures against COVID-19, including numerous vaccine candidates at Canadian universities, such as the International Vaccine Centre (VIDO-InterVac), University of Saskatchewan, aiming to start human testing in 2021. In March, the Canadian government announced Can$192 million specifically for developing a COVID-19 vaccine, with plans to establish a national "vaccine bank" of several new vaccines that could be used if another coronavirus outbreak occurs. On 27 February 2020, a Generex subsidiary company, NuGenerex Immuno-Oncology is beginning a vaccine project to create an Ii-Key peptide vaccine against COVID-19. They wanted to produce a vaccine candidate that could be tested in humans "within 90 days."

The Jenner Institute of the University of Oxford announced that it had developed a vaccine candidate based on an adenovirus vector, and signed a manufacturing contract with Advent. They announced plans to start animal studies in March 2020, and human safety trials the following month. Washington University in St. Louis announced its efforts to develop a vaccine on 5 March 2020. On 5 March 2020, the United States Army Medical Research and Materiel Command at Fort Detrick and the Walter Reed Army Institute of Research in Silver Spring, both in western Maryland, announced they were working on a vaccine. Emergent Biosolutions announced that it had teamed with Novavax Inc. in the development and manufacture of a vaccine. The partners further announced plans for preclinical testing and a Phase I clinical trial by July 2020. On 12 March 2020, India's Health Ministry announced they are working with 11 isolates and that even on a fast track it would take at least around one-and-a-half to two years to develop a vaccine. On 12 March 2020, Medicago, a biotechnology company in Quebec City, Quebec, reported development of a coronavirus virus-like particle under partial funding from the Canadian Institutes for Health Research. The vaccine candidate is in laboratory research, with human testing planned for July or August 2020.

On 16 March 2020, the European Commission offered an €80 million investment in CureVac, a German biotechnology company, to develop a mRNA vaccine. Earlier that week, The Guardian had reported the US President Donald Trump offered CureVac "'large sums of money' for exclusive access to a Covid-19 vaccine", with the German government contesting this effort.

On 17 March 2020, American pharmaceutical company Pfizer announced a partnership with German company BioNTech to jointly develop a mRNA-based vaccine. mRNA-based vaccine candidate BNT162, currently in pre-clinical testing with clinical trials expected to begin in April 2020.

In Italy on 17 March 2020, Takis Biotech, an Italian biotech company announced they will have pre-clinical testing results in April 2020 and their final vaccine candidate could begin human testing by fall.

In France on 19 March 2020, the Coalition for Epidemic Preparedness Innovations (CEPI) announced a US$4.9 million investment in a COVID-19 vaccine research consortium involving the Institut Pasteur, Themis Bioscience (Vienna, Austria), and the University of Pittsburgh, bringing CEPI's total investment in COVID-19 vaccine development to US$29 million. CEPI's other investment partners for COVID-19 vaccine development are Moderna, Curevac, Inovio, Novavax, the University of Hong Kong, the University of Oxford, and the University of Queensland.

On 20 March 2020, Russian health officials said that scientists began tests with six different vaccine candidates in laboratory animal experiments.

Imperial College London researchers announced on 20 March 2020 that they are developing a self-amplifying RNA vaccine for COVID-19. The vaccine candidate was developed within 14 days of receiving the sequence from China.

Technology scale-up:

In March 2020, the US government, industry, and universities pooled resources to access supercomputers from IBM, combined with cloud computing resources from Amazon, Microsoft, and Google. The COVID-19

High Performance Computing Consortium is being used to forecast disease spread, model possible vaccines, and screen thousands of chemical compounds to design a COVID-19 vaccine or therapy.

Rumors and misinformation

Social media posts have promoted a conspiracy theory claiming the virus behind COVID-19 was known and that a vaccine was already available. The patents cited by various social media posts reference existing patents for genetic sequences and vaccines for other strains of coronavirus such as the SARS coronavirus.

- ## COVID-19 drug repurposing research:

Drug repositioning (also called drug repurposing) – the investigation of existing drugs for new therapeutic purposes – is a line of scientific research which aims to develop safe and effective COVID-19 treatments. Other research directions include the scientific development of a COVID-19 vaccine.

Several existing antiviral medications, previously developed or used as treatments for SARS, MERS, HIV, and malaria, are being researched as COVID-19 treatments, with some moving into clinical trials. In a statement to the journal Nature Biotechnology in February 2020, US NIH Viral Ecology Unit Chief Vincent Munster said, "The general genomic layout and the general replication kinetics and the biology of the MERS, SARS and [SARS-CoV-2] viruses are very similar, so testing drugs which target relatively generic parts of these coronaviruses is a logical step".

Drug repurposing

Drug repositioning (also known as drug repurposing, re-profiling, re-tasking or therapeutic switching) is the repurposing of an approved drug for the treatment of another disease or medical condition.

Reusing approved medications has the following benefits:

- Faster legislative and medical approval
- Existing body of scientific literature (particularly Contraindications and Drug Interactions)

Studies

Chloroquine

Chloroquine is an anti-malarial medication that was studied. On 18 March, the WHO announced that chloroquine and the related hydroxychloroquine would be among the four drugs studied as part of the Solidarity clinical trial. New York governor Andrew Cuomo announced that New York State trials of chloroquine and hydroxychloroquine would begin on 24 March.

Favipiravir

Chinese clinical trials in Wuhan and Shenzhen claimed to show that favipiravir, was "clearly effective". One Japanese public health official said the drug may be less effective in severe cases.

On 22 March 2020, Italy approved the drug for experimental use against COVID-19 and began conducting trials in the three regions most affected by the disease. The Italian Pharmaceutical Agency reminded the public that the existing evidence in support of the drug is scant and preliminary.

Interferon beta

A UK firm started conducting trials on IFN-β, a drug that was originally developed to treat COPD.

Lopinavir/ritonavir

One study of lopinavir/ritonavir (Keletra), a combination of the antivirals lopinavir and ritonavir, concluded that "no benefit was observed". The drugs were designed to inhibit HIV from replicating by binding to the protease. A team of researchers at the University of Colorado are trying to modify the drugs to find a compound that will bind with the protease of COVID-19.

There are criticisms within the scientific community about directing resources to repurposing drugs specifically developed for HIV/AIDS. The WHO included lopinavir/ritonavir in the international Solidarity trial.

Remdesivir

Cleveland University Hospitals started conducting two clinical trials for remdesivir, one for people with moderate illness and another for those with more severe illness. The Feinstein Institute of the Northwell Health system partnered with Gilead Sciences on phase III clinical trials for remdesivir.

Zithromax

New York State began trials for zithromax on 24 March.

Repurposed drugs

- **Antiviral drugs**

Since SARS-CoV-2 is a virus, considerable scientific attention has been focused on repurposing approved anti-viral drugs that were developed for prior outbreaks such as MERS, SARS, and West Nile Virus.

- **Anti malarial agents**

Chloroquine: Wang et al. have shown chloroquine to effectively inhibit 2019-nCoV in vitro.

Hydroxychloroquine

- **Broad-spectrum agents**

Ribavirin: ribavirin was recommended for COVID-19 treatment according to Chinese 7th edition guidelines

Umifenovir: umifenovir was recommended for COVID-19 treatment according to Chinese 7th edition guidelines

- **Interferons**

Interferon-β — Previously tested against MERS in combination with other drugs.

- **Drugs originally developed for SARS**

APN01 (ACE2 protein decoy)

- **Antibiotics**

Some antibiotics that have been identified as potentially repurposable as COVID-19 treatments

- Teicoplanin
- Oritavancin
- Dalbavancin
- Monensin
- Emetine

Anti-cytokine storm

Cytokine storm, a life-threatening medical condition, can be a complication in the later stages of severe COVID-19. There is evidence that hydroxychloroquine has anti-cytokine storm properties.

Tocilizumab has been included in treatment guidelines by China's National Health Commission after a small study was completed. It is undergoing a phase 2 non randomized test at the national level in Italy after showing positive results in people with severe disease. Combined with a serum ferritin blood test to identify cytokine storms, it is meant to counter such developments, which are thought to be the cause of death in some affected people. The interleukin-6 receptor antagonist was approved by the FDA for treatment against cytokine release syndrome induced by a different cause, CAR T cell therapy, in 2017.

The Feinstein Institute of Northwell Health announced in March a study on "a human antibody that may prevent the activity" of IL-6.

Enhanced innate immunity

The eicosanoid signaling molecule leukotriene B4 (LTB4) is a possible (as yet untested) candidate drug for treatment of COVID-19. In the lungs, LTB4 is produced in alveolar macrophages, and it recruits neutrophil white blood cells. A 1989 study in humans reported an approximately 17-fold (1700%) increase in neutrophil white blood cells in recovered human lung fluid 4 hours after LTB4 injection into the human lung (10 ml LTB4 diluted to 5 x 10-7 M in sterile saline). Notably, there was no significant change in lung epithelial permeability associated with LTB4-induced neutrophil recruitment. Therefore, it was suggested that LTB4 can recruit neutrophils without initiating an inflammatory cascade within the lungs.

Several subsequent studies have reported that LTB4 exerts potent anti-viral (and general anti-microbial) effects in the lung. For example, a 2008 study in mice reported that daily i.v. treatment with LTB4 led to a significant decrease in lung viral loads of influenza A at day 5 post-infection. A follow-up study in 2009 reported that LTB4 potentiates cytokine signaling by human white blood cells (neutrophils in particular), and that this action is not sensitive to chloroquine (raising the possibility that chloroquine and LTB4 could be used in combination to combat viral infection).

More recently, a 2019 study in mice reported that a single dose of LTB4 as late as five days post-infection with influenza A virus (at peak viral load) reduces proliferation of inflammatory monocyte-derived macrophages, controls tissue damage and enhanced significant survival against lethal influenza A infection. With regard to the mechanism of LTB4 action, it was found that LTB4 promotes production of interferon-α (IFN-α) (involved in innate immunity in viral infection) by pulmonary interstitial macrophages, resulting in inhibition of inflammatory monocyte-derived macrophages (the excessive accumulation of which is associated with 'cytokine storm' and mortality). However, in influenza A infected mice, LTB4 did not significantly alter neutrophil number.

Furthermore, a 2011 study in humans reported that nasal fluids recovered from healthy subjects who received nasally-administered LTB4, efficiently killed human coronavirus, respiratory syncytial virus, and influenza B virus.

Passive antibody therapy:

Transfering donated blood containing antibodies produced by the immune systems of those who have recovered from COVID-19 to people who need them is being investigated as a nonvaccine method of immunization. This strategy was tried for SARS. Viral neutralization is the anticipated mechanism of action by which passive antibody therapy can mediate defense against SARS-CoV-2. Other mechanisms, such as antibody-dependent cellular cytotoxicity and/or phagocytosis, may however be possible. Other forms of passive antibody therapy, for example, using manufactured monoclonal antibodies, are in development. Production of 'convalescent serum', which consists of the liquid portion of the blood from recovered patients and contains antibodies specific to this virus, could be increased for quicker deployment.

****

Chapter-5

Affected Countries and their related reports (till 23rd March of 2020)

Affected Countries and their related reports										
Country	**China**	**Italy**	**Spain**	**United States**	**Iran**	**Germany**	**France**	**South Korea**	**Switzerland**	**United Kingdom**
Confirmed Cases	81,093	59,138	33,089	33,073	23,049	22,672	16,018	8,961	7,014	5,683
REPORTED DEATHS	3,270	5,476	2,182	416	1,812	86	674	111	60	281
RECOVERED	61,644	1,045	183	8	2,959	25	12	288	4	19
Country	**Netherlands**	**Belgium**	**Austria**	**Norway**	**Sweden**	**Australia**	**Portugal**	**Malaysia**	**Canada**	**Denmark**
Confirmed Cases	4,749	3,401	3,244	2,132	1,906	1,609	1,600	1,518	1,430	1,420
REPORTED DEATHS	213	75	16	7	21	7	14	14	20	13
RECOVERED	-	1	4	1	1	21	-	26	8	1
Country	**Turkey**	**Brazil**	**Japan**	**Czech Republic**	**Israel**	**Luxembourg**	**Ireland**	**Thailand**	**Cruise ship (Diamond**	**Pakistan**
Confirmed Cases	1,236	1,128	1,055	1,047	883	798	785	721	712	646
REPORTED DEATHS	30	15	37	0	1	8	3	1	8	3
RECOVERED	-	-	118	-	4	-	-	34	-	2
Country	**Chile**	**Indonesia**	**Poland**	**Ecuador**	**Greece**	**Finland**	**Saudi Arabia**	**Singapore**	**Qatar**	**Romania**
Confirmed Cases	632	579	563	532	530	521	511	509	481	433

REPORTED DEATHS	1	49	7	7	13	1	0	2	0	2
RECOVERED	-	2	-	-	-	1	1	152	-	6

Country	Slovenia	Iceland	Philippines	Russia	India	Peru	Hong Kong	Panama	Bahrain	Estonia
Confirmed Cases	414	409	396	367	360	318	317	313	310	306
REPORTED DEATHS	1	1	33	1	7	5	4	3	1	0
RECOVERED	-	1	2	3	4	-	-	-	35	-

Country	Egypt	Mexico	Lebanon	South Africa	Argentina	Iraq	Colombia	Croatia	Armenia	Serbia
Confirmed Cases	294	251	248	240	225	214	210	206	190	188
REPORTED DEATHS	8	2	4	0	4	17	1	1	0	0
RECOVERED	27	4	1	-	-	15	-	-	-	-

Country	Slovakia	Kuwait	Taiwan	Bulgaria	San Marino	United Arab	Algeria	Uruguay	Hungary	Latvia
Confirmed Cases	178	176	169	163	160	153	139	135	131	124
REPORTED DEATHS	1	0	2	3	20	2	11	0	6	0
RECOVERED	-	2	17	-	-	17	-	-	-	1

Country	Vietnam	Costa Rica	Morocco	Lithuania	New Zealand	Jordan	Bosnia and	Andorra	Brunei	Cambodia
Confirmed Cases	121	117	109	105	102	99	93	88	88	86
REPORTED DEATHS	0	2	3	1	0	0	1	1	0	0

RECOVERED	16	-	-	-	-	-	-	-	-	1

Country	**Sri Lanka**	**North Macedon**	**Cyprus**	**Faroe Islands**	**Moldova**	**Albania**	**Belarus**	**Burkina Faso**	**Malta**	**Dominica n**
Confirmed Cases	86	85	84	80	80	76	76	75	73	72
REPORTED DEATHS	0	1	1	0	1	2	0	3	0	2
RECOVERED	1	-	-	-	-	-	3	-	-	-

Country	**Venezuel a**	**Azerbaij an**	**Tunisia**	**Palestine**	**Kazakhst an**	**Senegal**	**Oman**	**Guadelo upe**	**Trinidad and**	**Georgia**
Confirmed Cases	70	65	60	59	57	56	52	51	50	49
REPORTED DEATHS	0	1	1	0	0	0	0	1	0	0
RECOVERED	-	3	-	-	-	1	9	-	-	-

Country	**Ukraine**	**Uzbekist an**	**Afghanis tan**	**Liechtens tein**	**Nigeria**	**Martiniq ue**	**Channel Islands**	**Réunion Island**	**Cameroo n**	**Guam**
Confirmed Cases	47	43	40	37	36	35	32	28	27	27
REPORTED DEATHS	3	0	1	0	1	1	0	0	0	1
RECOVERED	-	-	-	-	-	-	-	-	-	-

Country	**Banglade sh**	**Hondura s**	**Mauritiu s**	**Democra tic**	**Puerto Rico**	**Montene gro**	**Paraguay**	**Cruise ship**	**Cuba**	**Ghana**
Confirmed Cases	24	24	24	23	23	22	22	21	21	21
REPORTED DEATHS	2	0	2	1	1	1	1	0	1	1
RECOVERED	-	-	-	-	-	-	-	-	-	-

Country	Bolivia	Jamaica	Macau	French Guiana	Guatema la	Rwanda	Kenya	Barbados	Ivory Coast	Kyrgyzst an
Confirmed Cases	19	19	19	18	17	17	15	14	14	14
REPORTED DEATHS	0	1	0	0	1	0	0	0	0	0
RECOVERED	-	-	-	-	-	-	-	-	-	-

Country	Maldives	French Polynesia	Monaco	Gibraltar	Mongolia	Ethiopia	Togo	Guyana	Mayotte	Seychelle s
Confirmed Cases	13	11	11	10	10	9	9	7	7	7
REPORTED DEATHS	0	0	0	0	0	0	0	1	0	0
RECOVERED	-	-	-	-	-	-	-	-	-	-

Country	Equatori al	Tanzania	US Virgin	Aruba	Bahamas	Eswatini	Saint Martin	Cape Verde	Cayman Islands	Central African
Confirmed Cases	6	6	6	5	4	4	4	3	3	3
REPORTED DEATHS	0	0	0	0	0	0	0	0	1	0
RECOVERED	-	-	-	-	-	-	-	-	-	-

Country	Congo	Curacao	El Salvador	Gabon	Liberia	Madagas car	Namibia	Saint Barthele	Zambia	Zimbabw e
Confirmed Cases	3	3	3	3	3	3	3	3	3	3
REPORTED DEATHS	0	1	0	1	0	0	0	0	0	0
RECOVERED	-	-	-	-	-	-	-	-	-	-

Country	Angola	Benin	Bermuda	Bhutan	Fiji	Guinea	Haiti	Kosovo	Mauritania	New Caledoni
Confirmed Cases	2	2	2	2	2	2	2	2	2	2
REPORTED DEATHS	0	0	0	0	0	0	0	1	0	0
RECOVERED	-	-	-	-	-	-	-	-	-	-

Country	Nicaragua	Niger	Saint Lucia	Sudan	Suriname	Antigua and	Chad	Djibouti	Dominica	Eritrea
Confirmed Cases	2	2	2	2	2	1	1	1	1	1
REPORTED DEATHS	0	0	0	1	0	0	0	0	0	0
RECOVERED	-	-	-	-	-	-	-	-	-	-

Country	Gambia	Greenland	Grenada	Montserrat	Mozambique	Nepal	Saint Vincent and	Somalia	Syria	Timor-Leste
Confirmed Cases	1	1	1	1	1	1	1	1	1	1
REPORTED DEATHS	0	0	0	0	0	0	0	0	0	0
RECOVERED	-	-	-	-	-	1	-	-	-	-

Country	Uganda	Vatican City	TOTAL
Confirmed Cases	1	1	3,37,727
REPORTED DEATHS	0	0	15,140
RECOVERED	-	-	66,907

Source: WHO, CNA and Johns Hopkins University. Numbers of Recovered cases may not be exhaustive as not all countries report these statistics. (till 22nd March, 2020)

****

Chapter-6

History of the coronavirus pandemic

This chapter documents the chronology and epidemiology of severe acute respiratory syndrome coronavirus 2 (SARS-CoV-2), the virus responsible for the 2019–20 coronavirus pandemic originating in Wuhan, China. This chronological section covers November 2019 to 24th march 2020.

17 November 2019

A confirmed case of the novel coronavirus emerged on 17 November 2019, according to 13 March 2020 reports of official Chinese government sources but was not recognized at that time. There may have been earlier patients; the search for them continues.

1 December

The first known patient started experiencing symptoms on 1 December 2019. He had not been to the Huanan Seafood Wholesale Market of Wuhan. No epidemiological link could be found between this case and later cases.

8–18 December

Between 8 and 18 December 2019, seven cases later diagnosed with novel coronavirus were documented; two of them were linked with the Huanan Seafood Wholesale Market; five were not.

12 December

Chinese state broadcaster CCTV reported in a broadcast airing on 12 January 2020 that a "new viral outbreak was first detected in the city of Wuhan, China, on 12 December 2019."

21 December

Chinese epidemiologists with the Chinese Center for Disease Control and Prevention (CCDC) published an article on 20 January 2020 stating that the first cluster of patients with "pneumonia of an unknown cause" had been identified on 21 December 2019.

24 December

First publicly reported collection of virus sample from patient for genetic sequencing.

25 December

Report of medical staffs in two hospitals in Wuhan suspected to be infected and were being isolated around 25 December.

26 December

On 26 December 2019, a laboratory identified the coronavirus from the sample collected on 24 December as to be most closely related to a bat SARS-like coronavirus.

27 December

On 27 December 2019, almost complete genetic sequencing of the previous sample have been finished and shared to Institute of Pathogen Biology, Chinese Academy of Medical Sciences & Peking Union Medical College (CAMS&PUMC).

29 December

According to a CCDC publication on 31 January 2020, the facts leading up to the identification of the 2019-nCoV were as follows, "On 29 December 2019, a hospital in Wuhan admitted four individuals with pneumonia and recognized that all four had worked in the Huanan Seafood Wholesale Market, which sells live poultry, aquatic products, and several kinds of wild animals to the public. The hospital reported this occurrence to the CCDC, which led Wuhan CCDC staff to initiate a field investigation with a retrospective search for pneumonia patients potentially linked to the market. The investigators found additional patients linked to the market, and on 30 December, health authorities from Hubei Province reported this cluster to CCDC. The following day, CCDC sent experts to Wuhan to support the investigation and control effort. Samples from these patients were obtained for laboratory analyses."

30 December

On 30 December 2019, genetic sequencing report of the pathogen of a patient indicated inaccurately the discovery of Severe acute respiratory syndrome coronavirus (SARS coronavirus) in the test result. After receiving the test result, multiple doctors in Wuhan shared the information via internet, including Dr. Li Wenliang, an ophthalmologist at Wuhan Central Hospital, who posted a warning to alumni from his medical school class via a WeChat online forum that a cluster of seven patients treating within the ophthalmology department had been unsuccessfully treated for symptoms of viral pneumonia and diagnosed with SARS. Because these patients did not respond to traditional treatments, they were quarantined in an ER department of the Wuhan Central Hospital. In the WeChat post, Li falsely posted that "X Hospital has many confirmed cases of SARS" and "There had been 7 confirmed cases of SARS". Dr. Li posted a snippet of an RNA analysis finding "SARS coronavirus" and extensive bacteria colonies in a patient's airways. Dr. Li contracted this coronavirus from a patient he treated, was hospitalized on 12 January 2020 and died on 7 February 2020.

News of an outbreak of "pneumonia of unknown origin" started circulating on social media on the evening of 30 December 2019. The social media reports stated that 27 patients in Wuhan—most of them stall holders at the Huanan Seafood Market—had been treated for the mystery illness.

On the evening of 30 December 2019, an "urgent notice on the treatment of pneumonia of unknown cause" was issued by the Wuhan Municipal Health Committee on its Weibo social media account. It was reported that since the beginning of December, there had been "a successive series of patients with unexplained pneumonia"—27 suspected cases in total, seven of which were in critical condition and 18 were stable, two of which were on the verge of being discharged soon. The Wuhan Municipal Health Committee reported to the WHO that 27 people had been diagnosed with pneumonia of unknown cause. Most were stallholders from the Huanan Seafood Wholesale Market, seven of whom were in critical condition. The Wuhan Municipal Health Commission also made a public announcement regarding the situation.

Early investigations into the cause of the pneumonia ruled out seasonal flu, SARS, MERS and bird flu.

Hong Kong Secretary for Food and Health Sophia Chan Siu-chee announced after an urgent night-time meeting with officials and experts, "[any suspected cases] including the presentation of fever and acute respiratory illness or pneumonia, and travel history to Wuhan within 14 days before onset of symptoms, we will put the patients in isolation."

31 December

On 31 December 2019, China contacts the WHO and informs them of "cases of pneumonia of unknown etiology (unknown cause) detected in Wuhan"

An "urgent notice on the treatment of pneumonia of unknown cause" was issued to the Wuhan Municipal Health Center.

As a result of the official announcement of the Wuhan Municipal Health Commission, Hong Kong, Macau and Taiwan immediately tightened their inbound screening processes.

Qu Shiqian, a vendor at the Huanan Seafood Market, said government officials had disinfected the premises on 31 December 2019 and told stallholders to wear masks.

"Chinese state television reported that a team of experts from the National Health Commission had arrived in Wuhan on 31 December 2019 to lead the investigation, while the People's Daily said the exact cause remained unclear and it would be premature to speculate." Chinese state broadcaster CCTV reported that a team of senior health experts had been dispatched to the city of Wuhan and were reported to be "conducting relevant inspection and verification work."

Tao Lina, a public health expert and former official with Shanghai's Centre for disease control and prevention, said, "I think we are quite capable of killing it in the beginning phase, given China's disease control system, emergency handling capacity and clinical medicine support."

1 January 2020

According to information reported by the South China Morning Post on 13 March 2020, Chinese authorities had identified 266 people who had been infected before the beginning of 2020.

According to the Chinese state-sponsored Xinhua News, the Huanan Seafood Market was closed on 1 January 2020 for "regulation." However, in the Consortium's report of 24 January 2020, it was stated that the Huanan Seafood Market had been closed on 1 January 2020 for "cleaning and disinfection. However the virus could only stay on surfaces for so long so this was useless"

2 January

On 2 January, 41 admitted hospital patients in Wuhan, China, were confirmed to have contracted (laboratory-confirmed) the 2019-nCoV (Novel coronavirus); 27 (66%) patients had direct exposure to Huanan Seafood

Wholesale Market. All 41 patients were subsequently relocated from the hospital they had originally been diagnosed in to the Jinyintan Hospital in Wuhan, China.

3 January

On 3 January 2020, Chinese scientists at the National Institute of Viral Disease Control and Prevention (IVDC) determined the genetic sequence of the novel β-genus coronaviruses (naming it '2019-nCoV') from specimens collected from patients in Wuhan, China, and three distinct strains were established.

Health authorities in Wuhan reported 44 cases, a big jump from the 27 reported on Tuesday. Eleven of the 44 were seriously ill, the Wuhan Municipal Health Commission said, although there had been no reported deaths to date. The health of the 121 close contacts of the cases was being monitored.

On 3 January 2020, Dr. Li Wenliang, a Wuhan ophthalmologist, was summoned to the Wuhan Public Security Bureau where he was told to sign an official confession and admonition letter promising to cease spreading false "rumors" regarding the coronavirus. In the letter, he was accused of "making false comments" that had "severely disturbed the social order". The letter stated, "We solemnly warn you: If you keep being stubborn, with such impertinence, and continue this illegal activity, you will be brought to justice—is that understood?" Dr. Li signed the confession writing: "Yes, I understand."

In late January, The Supreme People's Court rebuked police for punishing Li and his fellow doctors. The Wuhan police clarified that Li Wenliang was not arrested or fined, but was warned as he had spread that "There had been 7 confirmed cases of SARS", which was not true.

The HHS Secretary Dr. Alex Azar was alerted on January 3 that CDC director Dr. Robert Redfield had discussions with Chinese doctors about the virus.

4 January

The head of the University of Hong Kong's Centre for Infection, Ho Pak-leung, warned that the city should implement the strictest possible monitoring system for a mystery new viral pneumonia that infected dozens of people on the mainland, as it was highly possible that the illness was spreading from human to human. The microbiologist also warned that there could be a surge in cases during the upcoming Chinese New Year. Ho said he hoped the mainland would release more details as soon as possible about the patients infected with the disease, such as their medical history, to help experts analyse the illness and to allow for more effective preventive measures to be put in place.

The Singapore Ministry of Health said on Saturday, 4 January, that it had been notified of the first suspected case of the "mystery Wuhan virus" in Singapore, involving a three-year-old girl from China who had pneumonia and a travel history to the Chinese city of Wuhan. On 5 January, the Singapore Ministry of Health released a press statement stating that the earlier suspected case was not linked to the pneumonia cluster in Wuhan and was also tested negative for the SARS and MERS-CoV.

Chinese officials were criticized for failing to disclose any information about the "mysterious virus" that machine translations of official reports suggested may be caused by a new coronavirus.

The WHO waited for China to release information about the "mysterious new pneumonia virus". The United Nations agency activated its incident-management system at the country, regional and global level and was standing ready to launch a broader response if it was needed. The WHO's regional office in Manila said in Twitter posts Saturday.: "China has reported to WHO regarding a cluster of pneumonia cases in Wuhan, Hubei Province.The Govt has also met with our country office, and updated @WHO on the situation. Govt actions to control the incident have been instituted and investigations into the cause are ongoing."

The Wuhan Institute of Virology didn't respond to an emailed request for comment on the infectious source.

5 January

The number of suspected cases reached 59 with seven in a critical condition. All were quarantined and local medical officials commenced the monitoring of 163 of their contacts. At this time, there had been no reported cases of human-to-human transmission or presentations in healthcare workers.

6 January

On Monday, 6 January, the Wuhan health authorities announced they continued seeking the cause but had so far ruled out influenza, avian influenza, adenovirus, and coronaviruses SARS and MERS as the respiratory pathogen that had infected 59 people as of 5 January.

7 January

Since the outburst of social media discussion of the mysterious pneumonia outbreak in Wuhan, China, Chinese authorities censored the hashtag WuhanSARS and were now investigating anyone who was allegedly spreading misleading information about the outbreak on social media.

The world continued to wait for China to disclose more information about what had triggered an unexplained pneumonia outbreak in Wuhan, China's tenth-largest city.

"The U.S. Centers for Disease Control and Prevention (CDC) issued a travel notice Monday for travelers to Wuhan, Hubei province, China due to the cluster of cases of pneumonia of an unknown etiology..."

8 January

Scientists in China announced the discovery of a new coronavirus.

South Korea announced the first possible case of virus coming from China. South Korea put a 36-year-old Chinese woman under isolated treatment amid concerns that she had brought back a form of viral pneumonia that had sickened dozens in mainland China and Hong Kong in the previous weeks. The unidentified woman, who worked for a South Korean company near capital Seoul, had experienced cough and fever since returning from a five-day trip to China on 30 December, the KCDC said in a press release. The woman had spent time in Wuhan, China, but had not visited the Huanan Seafood Market.

9 January

The WHO confirmed that the novel coronavirus had been isolated from one person who had been hospitalised. On the same day, the European Centre for Disease Prevention and Control posted its first risk assessment. The WHO also reported that Chinese authorities had acted swiftly, identifying the novel coronavirus within weeks of the onset of the outbreak, with the total number of positively tested people being 41. The first death from the virus occurred in a 61-year-old man who was a regular customer at the market. He had several significant medical conditions, including chronic liver disease, and died from heart failure and pneumonia. The incident was reported in China by the health commission via Chinese state media on 11 January.

Chinese scientists reported on Chinese state broadcaster CCTV that they had found a new "coronavirus in 15 of 57 patients with the illness in the central city of Wuhan, saying it has been preliminarily identified as the pathogen for the outbreak". The scientists announced that the current 'Wuhan Virus', a coronavirus, appears to not be as lethal as SARS. They reported that the new viral outbreak was first detected in the city of Wuhan on 12 December 2019. Additionally, a total of 59 people have been identified as contracting the illness, seven patients had been in a critical condition at some stage, and no healthcare workers were reported as having been infected.

10 January

The gene sequencing data of the isolated 2019-nCoV, a virus from the same family as the SARS coronavirus, was posted on Virological.org by researchers from Fudan University, Shanghai. A further three sequences from the Chinese Center for Disease Control and Prevention, one from the Chinese Academy of Medical Sciences, and one from Jinyintan Hospital in Wuhan were posted to the Global Initiative on Sharing All Influenza Data (GISAID) portal. The same day, Public Health England issued its guidance.

On 10 January 2020, Dr. Li Wenliang, Chinese ophthalmologist and coronavirus whistleblower, started having symptoms of a dry cough. On 12 January 2020, Dr. Wenliang started having a fever. He was admitted to the hospital on 14 January 2020. His parents also contracted the coronavirus (presumably from Dr. Wenliang) and were admitted to the hospital with him. Dr. Wenliang tested negative several times for the coronavirus until finally testing positive on 30 January 2020.

First two patients in Shenzhen, Guangdong, China attend University of Hong Kong-Shenzhen Hospital.

11 January

The first two patients in Shenzhen city transferred into negative pressure room in Third People's Hospital of Shenzhen City due to matching lab test result, symptoms, and epidemiology and are being listed as suspected cases. The cases were not confirmed at the time, because requirement from the Chinese government at the time was that first case in each individual cities need to be submitted to provincial CDC, verified by national CDC, and then evaluated and confirmed by a specific diagnostic team in national CDC.

11–12 January

In China, more than 700 close contacts of the 41 confirmed cases, including more than 400 healthcare workers, had been monitored, with no new cases reported in China since 5 January. The WHO published initial guidance on travel advice, testing in the laboratory and medical investigation.

13 January

The USCDC announced that the genome had been posted on the NIH genetic sequence database, GenBank. On the same day, Thailand witnessed the first confirmed case of 2019-nCoV, the first outside China. The affected 61-year-old Chinese woman, who is a resident of Wuhan, had not visited the Huanan Seafood Wholesale Market, but was noted to have been to other markets. She had arrived in Bangkok on 8 January.

14 January

On 14 January, two of the 41 confirmed cases in Wuhan were reported to include a married couple, raising the possibility of human-to-human transmission.

On 14 January, Maria Van Kerkhove, acting head of WHO's emerging diseases unit said that there had been limited human-to-human transmission of the coronavirus, mainly small clusters in families, adding that "it is very clear right now that we have no sustained human-to-human transmission"

15 January

A second death occurred in a 69-year-old man in China on 15 January. The WHO published a protocol on diagnostic testing for 2019-nCoV, developed by a virology team from Charité Hospital.

16 January

On 16 January, the WHO was alerted by Japan's Ministry of Health, Labour and Welfare that the first case in Japan, a 30-year-old male Chinese national had tested positive to 2019-nCoV during a hospital stay between 10 and 15 January. He had not visited the Huanan Seafood Wholesale Market, but possibly had close contact with an affected person in Wuhan.

17 January

On 17 January, Thailand's second confirmed case was reported in a 74-year-old woman who arrived in Bangkok on a flight from Wuhan. The number of laboratory-confirmed cases rose to 45 in China.

Yang Xiaobo, head of the Assets Supervision and Administration Commission, died of pneumonia caused by the virus on 17 January.

18 January

After the first 41 laboratory-confirmed cases were identified on January 2, 2020, Chinese officials announced no new cases for the next 16 days, then reported 17 additional laboratory-confirmed cases, three of which were in critical condition. This brought the number of laboratory-confirmed cases in China to 62. The patients' ages ranged from 30 to 79. 19 were discharged and eight remain critical.

On the same day, the Wuhan City government held an annual banquet in the Baibuting community celebrating the Chinese New Year with forty thousand families in attendance despite the officials' knowledge of the spread of the Wuhan coronavirus. They shared meals, plates and ate together. On 21 January 2020 when Wuhan mayor Zhou Xianwang was asked on state television why this banquet was held even after the number of cases had risen to 312 he responded, "The reason why the Baibuting community continued to host the banquet this year was based on the previous judgment that the spread of the epidemic was limited between humans, so there was not enough warning."

19 January

On 19 January, the first confirmed cases were reported in China, outside Wuhan, one in the southern province of Guangdong and two in Beijing. Wuhan reported 136 additional laboratory-confirmed cases, bringing the total number of laboratory-confirmed cases in China to 201. A new death was also reported in Wuhan, bringing the total number of fatalities in China to three.

20 January

On 20 January, after two medical staff were infected in Guangdong, China announced that the virus was human-to-human transmissible.

Scientists from the China CDC identified three different strains of the 2019-nCoV confirming that the original Wuhan coronavirus had mutated into two additional strains.

Chinese premier Li Keqiang urged decisive and effective efforts to prevent and control the epidemic. First confirmed case reported in South Korea. Beijing and Guangdong reported an additional three and thirteen laboratory-confirmed cases, respectively. Shanghai confirms its first case, bringing the total number of laboratory-confirmed cases in China to 218. The investigation team from China's National Health Commission confirmed for the first time that the coronavirus can be transmitted between humans. At least two people had become infected whilst living hundreds of miles from Wuhan.

Five attendees of an as-yet-unnamed private international sales company meeting of 109 attendees, 94 from overseas, held from 20–22 January at the Grand Hyatt Hotel, Singapore were diagnosed with the Wuhan coronavirus upon returning home: one from Malaysia, two from South Korea and two from Singapore. One of the attendees was from Wuhan, China. It was reported that the company held a buffet for their delegates. These four diagnoses were not reported until 5 February 2020. The first laboratory-confirmed case in Singapore of an unrelated 67 year-old native of Wuhan was not reported until 23 January 2020. These cases linked to the meeting were the first evidence that the Wuhan coronavirus had spread through human-to-human contact outside China, which the WHO has said is deeply concerning and could signal evidence of a much larger outbreak. As of 5 February 2020, the sister of a Malaysian who attended the meeting had been infected and four more local staff in Singapore were confirmed as having virus symptoms.

21 January

A total of 291 cases have now been reported across major cities in China, including Beijing and Shanghai. However, most patients are in Wuhan, the central city of 11 million at the heart of the outbreak.

A report by the MRC Centre for Global Infectious Disease Analysis at Imperial College London suggested there could be more than 1,700 infections. However, Gabriel Leung, the dean of medicine at the University of Hong Kong, put the figure closer to 1,300.

After 300 confirmed diagnoses and 6 deaths, Chinese state media warned lower-level officials not to cover up the spread of a new coronavirus. Officials declared that anyone who concealed new cases would "be nailed on the pillar of shame for eternity", the political body responsible for law and order said. Local Chinese officials initially withheld information about the epidemic from the public. It later vastly under-reported the number of people that had been infected, downplayed the risks and failed to provide timely information that experts say could have saved lives. In its commentary published online on Tuesday January 21, 2020, the Communist Party's Central Political and Legal Commission talked of China having learned a "painful lesson" from the SARS epidemic and called for the public to be kept informed. Deception, it warned, could "turn a controllable natural disaster into a man-made disaster".

The Wuhan Municipal Health Commission reported at least 15 medical workers in Wuhan have also been infected with the virus, with one in a critical condition.

Confirmed cases were reported in several new locations in China. Zhejiang province and Tianjin reported five and two laboratory-confirmed cases, respectively. Guangdong reported three additional laboratory-confirmed cases. Shanghai and Henan province reported an additional four and one laboratory-confirmed cases, respectively. One laboratory-confirmed case was reported in Sichuan province, and Chongqing reported five laboratory-confirmed cases. Shandong, Hunan, and Yunnan all reported one laboratory-confirmed case each. Jiangxi reported two laboratory-confirmed cases. The total number of laboratory-confirmed cases in China increased to 312 and the death toll increased to six.

New cases were also reported outside of mainland China. Taiwan reported its first laboratory-confirmed case, and the United States reported its first laboratory-confirmed case in the state of Washington, the first in North America.

China's Wuhan Institute filed to patent the use of Gilead's remdesivir for the treatment of novel coronavirus.

January 22.

'Health Commission: 440 cases of pneumonitis infected with new coronavirus have been diagnosed' - video news report from China News Service, 22 January 2020.

'Infectious Disease Expert Discusses Coronavirus Threat with VOA' - Video news report from Voice of America with Dr.Anthony Fauci, 22 January 2020

New cases: Macau and Hong Kong reported their first laboratory-confirmed cases, with Hong Kong reporting its second on the evening of 22 January. Beijing reported an additional five laboratory-confirmed cases, while Guangdong reported an additional nine laboratory-confirmed cases. Shanghai reported an additional five laboratory-confirmed cases, while Tianjin reported an additional two laboratory-confirmed cases. Zhejiang and Jiangxi reported an additional five and one laboratory-confirmed cases, respectively. Liaoning reported its first two laboratory-confirmed cases. Guizhou, Fujian, Anhui, Shanxi and Ningxia reported one laboratory-confirmed case each. Hainan reported four laboratory-confirmed cases. Hunan reported three additional laboratory-confirmed cases. Guangxi reported two laboratory-confirmed cases. In all, the total number of laboratory-confirmed cases in China increased to 571 and the death toll to 17.

Internationally, two more laboratory-confirmed cases were reported in Thailand, raising the total number of laboratory-confirmed cases in Thailand to four.

New data showed indications of the current rapid spread of the disease and an increase in the rate of transmission.

Officials announce a quarantine of the greater Wuhan, China area to commence 23 January 2020 at 10:00 am. No traffic will be allowed in or out of the city.

23 January

Jiangsu reported its first laboratory-confirmed case. Heilongjiang reported its first two laboratory-confirmed cases. Both Fujian and Guangxi reported an additional three laboratory-confirmed cases each. Shanghai reported an additional seven laboratory-confirmed cases. Xinjiang reported two laboratory-confirmed cases. Shaanxi reported three laboratory-confirmed cases. Gansu reported two laboratory-confirmed cases. Macau also reported its second laboratory-confirmed case, another 66-year-old man from Wuhan.

In all, the total number of laboratory-confirmed cases in mainland China increased to 628 while the death toll remained at 17.

Singapore reported its first laboratory-confirmed case, a 66-year-old man from China. Vietnam confirmed its first two laboratory-confirmed cases, a 65 or 66-year-old father and 27 or 28-year-old son from China.

A scientific preprint from the Wuhan institute of Virology is posted on Biorxiv announcing that a bat virus with 96% similarity had been sequenced in a Yunnan cave in 2013, whose sequence is posted the next day on public databases. It is confirmed by comparing infectivity of cells expressing or not expressing ACE2 that the novel coronavirus uses this same entry receptor as SARS-CoV.

Wuhan suspended all public transportation from 10 a.m. onwards, including all bus, metro and ferry lines. Additionally, all outbound trains and flights were halted.

24 January

'Hong Kong confirms two new cases of pneumonia' - video news report from China News Service, January 24, 2020.

Shandong reported six additional laboratory-confirmed cases. Hunan reported 15 additional laboratory-confirmed cases. Liaoning reported one additional laboratory-confirmed case. Fujian reported four additional laboratory-confirmed cases. Anhui reported six additional laboratory-confirmed cases. Ningxia reported one additional laboratory-confirmed case. Shanghai reported 13 additional laboratory-confirmed cases, bringing the total up to 33.

Japan, South Korea, and the United States all confirmed their second cases. Singapore confirmed its second and third cases. Thailand confirmed its fifth case. Hong Kong confirmed three additional cases, bringing the total number to five. Nepal confirmed its first case, a student who returned from Wuhan. France reported its first three confirmed cases, the first occurrences in the EU. The French Health Minister Agnès Buzyn stated that it is likely other cases would arise in the country.

The first confirmed incidence of human-to-human transmission outside of China was documented by the WHO in Vietnam.

A study by Chinese researchers indicates that people can be symptom-free for several days while the coronavirus is incubating, increasing the risk of contagious infection without forewarning signs.

A consortium of Chinese medical experts charged by the Chinese CDC with investigating the inception of the virus published their report in The Lancet, reporting details of 41 first known patients.

By the end of the day, the entire Hubei province had gone under a city-by-city quarantine, apart from Xiangyang and Shennongjia Forestry District.

25 January

Chinese Communist Party general secretary Xi Jinping called the "accelerating spread" of the coronavirus a "grave situation" in a Party Politburo meeting, and that it was "mutating" as Beijing escalates measures to contain the illness.

Australia confirmed its first four cases, one in Victoria and three in New South Wales. Malaysia reported its first three cases in Johor Bahru, and a fourth case later. Japan confirmed its third case. Canada confirmed its first case in Toronto. Thailand added two new cases for a total of seven. Singapore confirmed their fourth case.

A Chinese and a Sri Lankan suspected with the infection were admitted to a hospital in Sri Lanka.

Liang Wudong, a 62-year-old doctor, reportedly died in Hubei province from the coronavirus.

26 January

1000+ confirmed cases

100-999 confirmed cases

10-99 confirmed cases

1-9 confirmed cases

The Spring Festival holiday was extended to contain the coronavirus outbreak.

Shanghai reported its first death, an 88-year-old man.

The United States confirmed its third, fourth, and fifth cases: two in California and one in Arizona. Macau confirmed three additional cases, bringing its total to five. Hong Kong confirmed its sixth, seventh, and eighth cases. Thailand has confirmed its eighth case. The first of five patients was already discharged. There are another 39 suspected cases awaiting confirmation.

The Chinese Center for Disease Control and Prevention (CCDC) has started developing vaccines against the coronavirus, an official with the center said on Sunday.

Health officials in Ivory Coast are dealing with a suspected case of coronavirus, the country's health ministry has announced.

The United Nation's WHO Director-General Tedros Ghebreyesus said he was on his way to Beijing to confer with Chinese officials and health experts about the coronavirus outbreak.

China started requiring nationwide use of monitoring stations for screening, identification and immediate isolation of coronavirus-infected travellers, including at airports, railway stations, bus stations and ports.

A tentative clinical profile for the new coronavirus (2019-nCoV) was published by an assistant professor of population health science at the Icahn School of Medicine at Mount Sinai in New York. The lethality of the virus is unknown; however, the death toll has now climbed to above three percent.

Wang Xianliang, a Hubei provincial government official, died of pneumonia caused by the virus.

27 January

Dr. Gabriel Leung, Dean of the University of Hong Kong medical school and one of the foremost world experts on SARS and viruses, gave a three-hour presentation published on YouTube wherein he made nowcasts and forecasts of the coronavirus. Using traditional scientific modeling techniques that predict the spread of viruses, Dr. Leung projected the true number of coronavirus infections was likely 10 time more than the official reported numbers. Dr. Leung estimated that there were between 44,000–100,000 infections in China as of 24 January 2020. He stated that draconian measures were needed to slow the progress of the virus but that these measures would have no effect in stopping the coronavirus pandemic. He projected that the number of infections would continue exponentially peaking out in late April or May 2020. Dr. Leung predicted that at the peak of the pandemic, there could be up to 100,000 new infections per day. Dr. Leung subsequently published an article in The Lancet nowcasting and forecasting the likely progression of the Wuhan coronavirus taking into consideration numerous variables. Zhou Xianwang, the mayor of Wuhan, said on a Chinese state television talkshow that rules imposed by Beijing limited what he could disclose about the threat posed by the Wuhan coronavirus as it unfolded, suggesting "the central government was partially responsible for a lack of transparency that has marred the response to the fast-expanding health crisis."

Canada reported its first confirmed case and another presumptive case. Health officials have confirmed the fifth case of coronavirus in Australia, and have suspected an additional 5. The Sri Lankan Health Ministry confirms its first case of coronavirus, a 43 year old Chinese woman. Cambodia confirms its first case of the virus, a Chinese man who came with his family to Sihanoukville. Singapore confirms a fifth case, a 56 year old Chinese national who arrived from Wuhan on 18 January. Germany confirmed its first case in Bavaria, a case of domestic transmission. Taiwan reports its first case of domestic transmission of the coronavirus.

Beijing reports its first death from coronavirus.

Three new suspected cases in Austria; previous suspected cases tested negative. The 'Matei Balş' Institute reported the first possible case in Romania., but the first case would not be confirmed until 26 February. Ecuador reported a suspected case of coronavirus, a Chinese citizen who arrived from Hong Kong., but the first case would not be confirmed until 29 February. Fiji authorities are holding six Chinese travelers in quarantine in Nadi as a precaution after they failed to gain entry to Samoa due to Samoa's quarantine requirements that were implemented Friday. The quarantine requirements, imposed after an emergency Cabinet meeting, compel anyone who's been in China to "self-quarantine" in a country free from the coronavirus for 14 days. In Poland, two children were admitted to the Kraków hospital with the suspicion of coronavirus. In Mongolia a 14-year-old girl who was studying in China had fallen ill with a suspected case of pneumonia and laryngitis; she was pronounced dead on the same day. Health authorities have since taken a sample from the deceased girl to be analysed at the National Center for Communicable Diseases in Ulaanbaatar. Two Mongolian students returning from Taiwan to Chinggis Khaan International

Airport have shown symptoms of high fever and rising temperature and were put into quarantine after landing in Mongolia. In Switzerland, two people were put under quarantine at the Triemli Hospital in Zurich; both had previously been to China. These cases later turned up negative.

In Germany, the first specific, global case of coronavirus being transmitted by a person with no symptoms has been reported. The originally-infected individual is from Shanghai.

South Korean health officials met with Korean life science company representatives to speed the development of a test.

28 January

China's Supreme People's Court ruled that whistleblower, Li Wenliang, had not committed the crime of spreading "rumors" when on 30 December 2019 he posted to a WeChat forum for medical school alumni that seven patients under his care appeared to have contracted SARS. In their ruling, the Supreme People's Court stated, "If society had at the time believed those 'rumours', and wore masks, used disinfectant and avoided going to the wildlife market as if there were a SARS outbreak, perhaps it would've meant we could better control the coronavirus today," the court said. "Rumours end when there is openness."

Thailand confirms six more cases, bringing the total infected there to 14. Thailand's health minister, Anutin Charnvirakul, states that "we are not able to stop the spread" of coronavirus in the country. Singapore confirms two more cases, bringing the total infected in Singapore to seven. That was followed by a Hubei-related suspension from 29 January. Japan confirms 3 additional cases, bringing the total infected in Japan to seven, including a man who had never visited Wuhan. He was working as a tour bus driver and had driven a group from Wuhan earlier in January. Germany's first confirmed case, reported the previous day, had occurred in a German citizen who had not travelled to China. However, he had close contact with a visiting Chinese colleague who reported starting to feel ill during her return flight to Shanghai and she was diagnosed with coronavirus infection after arriving in China. Germany confirmed 3 new cases, all of whom were coworkers of the first confirmed patient. France confirmed its fourth case, an elderly Chinese tourist who is in critical condition.

The Brazilian Ministry of Health reports three suspected cases ongoing in three locations: Belo Horizonte (MG), Curitiba (PR) and São Leopoldo (RS). Canada reports a new presumptive case in British Columbia, a man in his 40s who had recently travelled to Wuhan.

A UK-Chinese medical research paper reports a statistical model finding that "estimates suggest the actual number of infected cases could be much higher than the reported, with estimated 26,701 cases (as of 28th January 2020)."

Scientists from The Peter Doherty Institute for Infection and Immunity (Doherty Institute) in Melbourne reported that they had successfully grown 2019-nCoV from a patient sample.

Xiangyang became quarantined starting 00:00; the entire Hubei province thus became quarantined save for Shennongjia Forestry District.

29 January

Tibet reported its first suspected case identified on the previous day and declared a level 1 health emergency in the evening, the last mainland provincial division to do so. Suspected cases have now been reported in all 31 mainland provincial divisions.

Companies in Hubei are required not to resume services before 13 February, and schools in Hubei are to postpone the reopening of schools.

The UAE confirms its first case. Shortly afterwards, an Emirates' news agency confirmed four people from a Chinese family to be infected. Finland reports its first case of the virus in Lapland, found in a Chinese tourist who left Wuhan before Wuhan was locked down. Singapore confirms three more cases of the virus, bringing the total infected to 10. Malaysia confirms three additional cases, bringing its total to seven. Japan reports four additional cases, including a tour bus guide that was on the same bus as one of the cases confirmed on 28 January and three evacuated from Wuhan. France confirmed a fifth case, the daughter of the patient in the fourth case.

Two Chinese nationals were placed in isolation wards in Armenia amid the first suspected case of coronavirus in the country. The Chinese nationals were tourists travelling to Armenia from neighbouring Georgia. Liana Torosyan, the head of the Department of Infectious Diseases, advised that samples will be sent to European labs, as Armenia does not have the capacity to test for the novel coronavirus. Brazil reports a total of 9 suspected cases in six states of the country.

Air Canada is halting all direct flights to China following the federal government's advisory to avoid non-essential travel to the mainland due to the coronavirus epidemic. The suspension is effective Thursday and slated to last until February 29.

30 January

Tibet confirms its first case, which was previously suspected. Cases have now been confirmed in all 31 provincial divisions of mainland China. India confirms its first case of coronavirus in a student who had returned from Wuhan University to the Indian state of Kerala. Philippines confirms its first case of coronavirus in a female Chinese national who arrived in Manila via Hong Kong on 21 January. Japan confirms three more cases, bringing the total to 14. Malaysia confirms one more case, bringing the total to eight. Singapore confirms three more cases, bringing the total to 13. South Korea confirms two more cases, with one of them being the first human-to-human transmission there. Vietnam confirms three new cases, bringing the total to five. France confirms its sixth case. Italy confirms its first two cases in a press conference by the Prime Minister, Giuseppe Conte. Germany confirms its fifth case, an employee of the company where the four previously known cases are also employed.

The United States confirmed its sixth case, the spouse of another patient in Chicago. This is the first confirmed case of human to human transmission within the United States.

31 January

The United Kingdom and Russia confirmed their first coronavirus infections. The first Swedish and Spanish cases were confirmed. The seventh confirmed case in the U.S. is in Santa Clara County, California. A fourth case of coronavirus in Canada has been confirmed in London, Ontario. Thailand confirmed five more cases with the first human-to-human virus transmission inside the country of a local taxi driver, bringing the total to 19. Singapore confirmed three more cases, bringing the total to 16. Chinese health experts warn the public that coronavirus patients can become reinfected. China starts repatriating citizens to Wuhan.

1 February

- Semi-log plot of cumulative incidence of confirmed cases and deaths in China and the rest of the world.

- Semi-log plot of daily incidence (epidemiology) of cases by region: Hubei Province; mainland China excluding Hubei; the rest of the world (ROW); and the world total.

- Semi-log plot of coronavirus daily deaths by region: Hubei Province; mainland China excluding Hubei; the rest of the world (ROW); and the world total.
- Australia reported another three cases, including the first two cases in South Australia, bringing their total to 12.
- Japan reported three more cases, increasing their total to 20.
- Singapore confirms two more cases, bringing the total to 18.
- South Korea confirmed the twelfth case of the coronavirus: a 49-year-old Chinese man who works as a Japanese tour guide in western Seoul.
- Spain confirmed its first case of the virus on La Gomera in the Canaries.
- The United States of America reported its eighth case, a man from Boston who recently returned to college after traveling to Wuhan.
- Vietnam confirmed its sixth case in Khánh Hòa Province, another domestic transmission in direct contact.

2 February

- India announced its second case.
- Philippines, saw the first confirmed death from COVID-19 outside mainland China announced. The case was that of a 44-year-old male who was the companion of the first confirmed case in the country; both are Chinese nationals from Wuhan who had arrived in the country via Hong Kong on 21 January. He had been in stable condition prior to his death on 1 February.
- South Korea reported three more cases, bringing their total to 15.
- United Arab Emirates reported its fifth case.
- The United States of America, confirmed three more cases, all in California, bringing the total to 11. Two Germans aboard the evacuation flight from Wuhan tested positive for the virus, bringing the number of cases there up to twelve.
- Vietnam announced its seventh case of COVID-19, a Vietnamese-American who had had a two-hour layover at Wuhan Airport.

3 February

- The number of deaths in Hubei outside Wuhan reached 101.
- According to WHO, there were no new countries reporting cases in the past 24 hours.
- India confirmed its third case in Kerala.
- Vietnam announced its eighth COVID-19 case, a Vietnamese woman who had been on the same flight with three other positive cases.

4 February

- Belgium confirmed its first case, bringing the total in the European Union to 24.
- Hong Kong confirmed its first death from a 39-year-old patient, the thirteenth confirmed case in the city.
- Malaysia confirmed two more COVID-19 cases, which included a Malaysian citizen, bringing the total to 10 cases.
- Singapore confirmed six more cases, including the first locally transmitted cases involving four, bringing the total to 24. 2 others came from an evacuation plane from Wuhan.
- South Korea confirmed its 16th COVID-19 case after a tourist returned from Thailand, the first to have been infected there.
- Thailand confirmed six more cases, bringing the total to 25.
- Vietnam announced its ninth and tenth COVID-19 case, including a Vietnamese man who had been on the same flight with four other positive cases.

5 February

- China, in Wuhan now reported 10,117 cases in total, exceeding 10,000. Tianjin, reported its first death. Chinese experts said that nucleic acid testing was only able to identify 30%-50% positive cases.
- Hong Kong confirmed three more cases, bringing the total to 21.
- Japan confirmed 10 new cases from the quarantined cruise ship Diamond Princess, near Yokohama, bringing the total number up to 35 cases. There were more than 3,500 on board to be tested.
- Malaysia announced two more cases, bringing the total cases to 12.
- Philippines confirmed its third positive case, a 60-year-old woman from Wuhan, China.

- Singapore confirmed four more cases, including a six-month-old Singaporean, bringing the total cases to 28.
- South Korea confirmed two more cases of COVID-19, including a patient who returned from Singapore. That brought the total number of cases to 18.
- The United States of America, saw Health officials in the state of Wisconsin announce the first case in that state.

6 February

- Canada announced 2 presumptive cases in British Columbia.
- China, In Zhejiang over 1,000 cases were confirmed, the first such province besides Hubei. Doctor Li Wenliang, a whistleblower who notified the public at large on social media of the coronavirus, died of the infection.
- Germany confirmed its thirteenth case, and Italy confirmed its third, bringing the total number of confirmed cases in Europe to 31.
- Hong Kong confirmed three more cases, bringing the total to 24.
- Japan confirmed another 10 new cases from Diamond Princess, quarantined in Yokohama, bringing the total to 45.
- Malaysia confirmed two more cases, bringing the total to 14. One of the cases was locally transmitted, the first in Malaysia.
- Singapore confirmed two more cases, bringing the total to 30.
- Taiwan announced three more cases, bringing the total to 16.
- The United Kingdom saw the third case confirmed, with the patient having visited Singapore.
- Vietnam confirmed two more cases, bringing the total to 12.

7 February

- Guangdong confirmed over 1,000 cases in total, the second such province outside Hubei, and recorded its first death.
- Wuhan reported 67 more deaths, bringing the total number of people killed by the virus there to 545; other cities in Hubei reported 14 deaths in total, their net figure now reaching 154.
- Gansu recorded its first death.
- Another 41 tested positive for the coronavirus on Diamond Princess, bringing the total number of cases in Japan to 86.
- Germany announced its fourteenth case.

- Hong Kong confirmed 2 more cases, bringing the total number to 26.
- Malaysia confirmed one more case from a tourist entering from Singapore, bringing the total to 15.
- Singapore confirmed three more cases, bringing the total to 33. Most of these cases were initially of unknown origin, causing the Disease Outbreak Response System Condition (DORSCON) level to increase to Orange.
- Vietnam confirmed one additional case, bringing the total to 13.

8 February

- China:

Wuhan confirmed 63 more deaths, bringing its total to 608; this now includes a Japanese citizen and a US citizen dying on this day.

Henan became the third province besides Hubei to report over 1,000 cases. Nationally, the number of cases outside Hubei exceeded 10,000.

Aerosol was confirmed as a medium of transmission for the first time.

Beijing reported a first type of case involving a patient, who tested positive after three negative test results, and Qingtian, Zhejiang reported a case where the patient tested positive only after being tested for the fifth time.

- France confirmed five cases involving British nationals, bringing the total number to 11.
- Japan, three more cases tested positive on Diamond Princess, bringing the total number of cases in Japan to 89.
- Malaysia confirmed one more case, bringing the total to 16.
- Singapore confirmed seven more cases, bringing the total number to 40.
- Thailand confirmed seven additional cases, bringing the total number to 32.
- United Arab Emirates confirmed two more cases, bringing the total number to seven.

9 February

- China (Mainland) now confirmed a total of 40,213 cases, exceeding 40,000, of which 29,631 were in Hubei. The coronavirus death toll in

China rose to 811, surpassing the toll from the SARS epidemic from 2002–03.

- Hong Kong confirmed ten more cases, with nine from the same family, bringing the total number to 36.
- Malaysia confirmed one more case, bringing the total number to 17.
- Singapore confirmed three more cases, bringing the total number to 43.
- South Korea reported three new cases of the virus, bringing the country's total to 27.
- Spain confirmed the second case in the country in Palma de Mallorca.
- Taiwan confirmed its eighteenth case.
- United Kingdom confirmed its fourth case.
- Vietnam confirmed its fourteenth case, a 55-year-old woman in Vĩnh Phúc.
- Six more cases were confirmed aboard Diamond Princess, bringing the total number of cases on the ship to 70, with Japan having 96 cases.

10 February

- Hong Kong confirmed six more cases, bringing the total number to 42.
- Japan confirmed 65 more cases on Diamond Princess, bringing the total to 135.
- Malaysia confirmed one more case involving a citizen, bringing the total to 18.
- Singapore confirmed another two cases including a Certis officer who served Quarantine Orders, bringing the total to 45.
- United Arab Emirates confirmed its eighth case, an Indian national.
- United Kingdom saw four additional cases were confirmed, bringing the total to eight. The transmission in these cases is believed to have occurred in France.
- United States of America confirmed its thirteenth case in San Diego, a patient who had been evacuated from Wuhan.

11 February

- Germany confirmed two more cases, bringing the total number to 16.
- Hong Kong confirmed seven more cases, bringing the total to 49.

- South Korea confirmed one more case, a 30-year-old Chinese woman, bringing the total number to 28.
- Singapore confirmed another two cases, bringing the total to 47.
- Vietnam confirmed its fifteenth case, a three-month-old baby infected by her grandmother.
- Thailand confirmed one new cases, bringing the total in the country to 33.
- WHO gives the disease the new name COVID-19. Additionally, the virus itself is named SARS-CoV-2.

12 February

- Hong Kong confirmed one more case, bringing the total number to 50.
- Japan confirmed 39 more cases on Diamond Princess, including one quarantine officer, bringing the total to 174. Another case was reported in Japan itself, bringing the total number to 29.
- Singapore confirmed three more cases, bringing the total to 50. Earlier, 300 employees of DBS Bank were asked to evacuate from the office at Marina Bay Financial Centre Tower 3 due to a confirmed case there.
- United Kingdom - The first case in London was confirmed, bringing the country's total to 9.
- United States of America confirmed one more case, bringing the total number to 14.

13 February

- Vietnam confirmed its 16th case.
- Japan confirmed four more cases, bringing the total to 33. At the same time, 44 new cases were confirmed on the Diamond Princess, bringing the total to 218. Later on, Japan confirmed its first death from the virus.
- Hong Kong confirmed three more cases, bringing the total number to 53.
- Malaysia confirmed one more case, bringing the total number to 19.
- Singapore confirmed eight more cases, bringing the total number to 58. A school in National University of Singapore will conduct e-learning from 14 to 21 February as a precaution after one of the cases involved a professor.

- Unites States of America, the CDC confirmed the 15th US coronavirus case, a Wuhan evacuee quarantined at a military base in Texas.
- The Chinese government has issued extension of order to shut down all non-essential companies, including manufacturing plants, in Hubei Province until at least 24:00 20 February 2020.
- Discrepancy in the number of cases The same day, Hubei reported newly confirmed cases of 14,840, nearly 10 times more than the previous day, while deaths more than doubled to 242. This was due to the change in definition that included clinical (radiological) diagnosis of patients. The World Health Organization indicated that for consistency it would report only the number of laboratory-confirmed cases.

14 February

- Canada confirmed British Columbia's fifth presumptive case, bringing the total number in the country to 8.
- Egypt confirmed its first case, a foreigner of undisclosed nationality. This was the first case on the African continent.
- Hong Kong confirmed three more infections, bringing the total number to 56.
- Japan confirmed four more cases, bringing the total number to 37.
- Singapore confirmed nine more cases, bringing the total number to 67.

15 February

- France saw the first death outside Asia being confirmed, an 80 year old Chinese tourist in France. The country also confirmed its 12th case.
- Japan confirmed nine additional confirmed cases not aboard Diamond Princess. Japan also confirmed 67 additional cases aboard the cruise ship Diamond Princess, bringing the total number of cases from the ship to 285.
- Malaysia confirmed three more cases, including an American passenger from the cruise ship MS Westerdam flying from Cambodia, bringing the total number to 22.
- Singapore confirmed five more cases, bringing the total number to 72.
- Thailand confirmed one more case, bringing the total number to 34.

16 February

- Hong Kong confirmed one new coronavirus case, bringing the total to 57.
- Japan also confirmed six new cases, bringing the total of infected people not aboard Diamond Princess to 59. 70 additional cases were confirmed on Diamond Princess, increasing the tally to 355.
- Singapore confirmed three more cases, bringing the total number to 75.
- South Korea confirmed one more case, bringing the total number to 29.
- Taiwan confirmed its first death from COVID-19, a man in his 60s. Moreover, two new cases were confirmed, bringing the total number to 20.
- United Arab Emirates confirmed one more case, a 37-year-old Chinese citizen, bringing the total number to 9.

17 February

- Hong Kong confirmed three more cases, bringing the total number to 60.
- Japan confirmed seven new cases, bringing the total of infected people not aboard Diamond Princess to 66. Diamond Princess in Japan confirmed 99 new coronavirus cases, bringing the total number of infected people on the cruise ship to 454.
- Singapore confirmed two more cases, bringing the total number to 77.
- South Korea confirmed one more case, bringing the total number to 30.
- Taiwan confirmed two more cases, bringing the total number to 22.
- Thailand confirmed one more case, bringing the total number to 35.

18 February

- Hong Kong confirmed two more cases, bringing the total number to 62.
- Japan confirmed eight more cases across the nation, bringing the total to 74. At the same time, 88 additional cases aboard Diamond Princess were confirmed, bringing the total on the ship to 542.
- Singapore confirmed four more cases, bringing the total number to 81.
- South Korea confirmed one more case, bringing the total number to 31.

19 February

- Hong Kong confirmed its second death from the coronavirus. In addition, three new cases were confirmed, bringing the total number in Hong Kong to 65.
- Iran confirmed its first two cases resulting in deaths.
- Singapore confirmed three more cases, bringing the total number to 84.
- South Korea confirmed 20 more cases, bringing the total number to 51 with the majority of these transmissions happening in a church in Daegu.
- Taiwan confirmed one more case, bringing the total number to 23.
- Passengers started to disembark from Diamond Princess, one of the cruise ships in the outbreak. The sheer number of passengers infected has led to questions over the effectiveness of quarantine measures. 79 more cases are confirmed on the cruise ship, bringing the total number to 621. Japan also reported 10 new cases not aboard the cruise ship.

20 February

- Hong Kong confirmed four more cases, bringing the total number to 69.
- Iran confirmed three more cases, bringing the total number to five.
- Japan confirmed ten more domestic cases, bringing the total number to 94.
- Singapore confirmed one more case, bringing the total number to 85.
- South Korea confirmed 53 more cases, bringing the total number to 104. The first death from the virus was also confirmed.
- Taiwan confirmed one more case, bringing the total number to 24.
- The United States confirmed one more case in California, bringing the total number to 16.
- Two deaths were confirmed aboard the cruise ship Diamond Princess along with 13 more cases, bringing the total number to 634.

21 February

- Australia confirmed four more cases involving evacuees from the cruise ship Diamond Princess, bringing the total number to 19.
- Canada confirms one more case, a woman who had recently visited Iran, bringing the total number to nine.

- Israel also confirmed its first case, an evacuee from Diamond Princess.
- Italy confirmed 17 cases, bringing the total number to 20. Authorities also reported the first death, a 78-year-old man.
- Iran announces 13 new cases, bringing the total to 18. Two more deaths were also confirmed.
- Japan confirmed 15 more cases, bringing the total number to 109.
- Lebanon confirmed its first case.
- Singapore confirmed one more case, bringing the total number to 86.
- South Korea confirmed 100 more cases, bringing the total number to 204. The country also reported its second death.
- Taiwan confirmed two more cases, bringing the total number to 26.
- United Arab Emirates confirmed two more cases, bringing the total number to 11.
- The United States confirmed 20 more cases, bringing the total number to 35. Furthermore, the Association of Public Health Laboratories (APHL) announced that only three states were capable of testing for the corona virus: California, Nebraska, and Illinois.

22 February

- Australia confirmed three more cases, bringing the total number to 22.
- Hong Kong confirmed one more case, bringing the total number to 70.
- Iran confirmed 10 more cases, bringing the total number to 28. A fifth death was also confirmed.
- Italy confirmed 59 more cases spread across three different administrative regions, bringing the total number to 79, making Italy the European country with the biggest number of cases of coronavirus infections. The second death was also confirmed, an Italian woman from Lombardy.
- Japan confirmed 26 more cases, bringing the total number to 135.
- Singapore confirmed three more cases, bringing the total number to 89.
- South Korea confirmed 229 more cases, bringing the total number to 433. Concerns are raised that 9 tourists from South Korea may have caused a widespread exposure in Israel.
- The United Arab Emirates confirmed two more cases, bringing the total number to 13.
- The urine sample of a patient tested positive.

- A genome analysis indicated that the virus in Huanan Seafood Market originated from outside.

23 February

- Canada confirmed one more presumptive case in Ontario. The samples had been sent to the National Microbiology Lab for further testing. Later, the cases was confirmed positive, bringing the total number to 10.
- Hong Kong confirmed four more cases, bringing the total number to 74.
- Iran confirmed 15 more cases and two more deaths, bringing the total number to 43 and 8 respectively.
- Israel confirmed one more case, who was a passenger on the cruise ship, bringing the total number to 2.
- Italy confirmed 73 new cases, bringing the total to 152 spread across five different administrative regions. The third death was also reported. Italy is now the third country in the world by number of cases, after China and South Korea.
- South Korea confirmed 169 more cases, bringing the total number to 602. Four more deaths were also confirmed.
- Taiwan confirmed two more cases, bringing the total number to 28.
- The United Kingdom confirmed four new cases, all involving evacuees from the Diamond Princess cruise ship, bringing the total number to 13.
- There were 57 more cases aboard Diamond Princess, bringing the total number to 691. One death associated to the cruise ship was also confirmed, bringing the total number to 3.

24 February

- Afghanistan confirmed its first case involving a person who recently returned from the Iranian city of Qom.
- Bahrain confirmed its first case involving a Bahraini citizen who travelled to Iran.
- Canada confirmed one more case in British Columbia, bringing the total number to 11.
- Hong Kong confirmed seven more cases, bringing the total number to 81.

- Iran confirmed 18 more cases, bringing the total number to 61. Four more deaths were also reported in the country, bringing the total number to 12.
- Iraq confirmed its first case involving an Iranian student.
- Italy confirmed 74 new cases, bringing the total to 229 spread across six different administrative regions. Four more deaths were also confirmed, bringing the total number to 7.
- Kuwait announced their first cases involving five people arriving from the Iranian city of Mashhad.
- Oman confirmed its first cases involving two Omani women who had come from Iran.
- Singapore confirmed one more case, bringing the total number to 90.
- South Korea confirmed 231 more cases, bringing the total number to 833. The seventh death was also confirmed.
- Spain confirmed its third positive case, an Italian man, in Tenerife, Canary Islands.
- Taiwan confirmed two more cases, bringing the total number to 30.
- The United States confirmed 18 more cases including evacuated passengers from the cruise ship, bringing the total number to 53.
- A patient in Sichuan province tested positive only on her 9th test.

25 February

- Algeria reported its first case, an Italian man who arrived on 17 February.
- Austria reported its first two cases, two Italians living in Tyrol tested positive for the coronavirus. One of the couple was a receptionist at a hotel and as a result the hotel in the Alpine tourist hub of Innsbruck was sealed off in an effort to contain the outbreak.
- Bahrain confirmed its second case, a Bahraini woman who had travelled from Iran through Dubai International Airport. This later increased to eight confirmed cases with the addition of two Bahraini men and four Saudi Arabian women. Later on the same day, Bahrain updated the number of cases to 17, with all infected people travelling from Iran. This total later rose again to 23 cases, again all having travelled from Iran.
- Brazil Ministry of Health reported the first positive case of coronavirus in the country and South America, a 61-year-old man

from São Paulo, who travelled to Lombardy, Italy, between 9 and 21 February. He was showing mild symptoms and quarantined at home, the confirmation test was also positive.

- Croatia announced its first case of the virus, with a patient hospitalised in the capital, Zagreb, who had travelled to Italy and stayed in Milan.
- France saw two additional cases: one was a French man returning from a trip in the Lombardy region of Italy; the other was a young Chinese woman returning to France from a trip to China. This took the total number of cases in the country to 14.
- Germany saw a case identified in the state of North Rhine-Westphalia of a person recently returned from Italy, taking the country's cases to 17. Another man was later confirmed to be positive, bringing the total number to 18.
- Hong Kong confirmed three more cases, bringing the total number to 84.
- Iranian officials confirmed the death toll was officially now 16. There were 34 more cases, bringing the total number to 95. The same day, Iranian MP Mahmoud Sadeghi confirmed he had tested positive for the virus, along with Iraj Harirchi, the Deputy Minister for Health, who had the previous day been part of a press conference about Iran's handling of the virus.
- Iraq confirmed five more cases - an Iranian student and a family of four who had arrived from Iran, taking the total number of cases to 6.
- Italy confirmed 94 new cases, bringing the total to 323 spread across nine different administrative regions. Three more deaths were also confirmed, bringing the total number to 10. This later in the day rose to four deaths, bringing the total to 11.
- Kuwait confirmed four additional cases, all of them people who had returned from Iran, taking their total to 9 cases.
- Oman confirmed two additional cases, taking its total to four cases. Both the additional cases were linked to travel to Iran.
- Singapore confirmed one more case, bringing the total number to 91.
- South Korea confirmed 144 more cases, bringing the total number to 977. Four more deaths were also confirmed, bringing the total to 11. Among those infected included a Korean Air crew member.
- Spain, Canary Islands: on the island of Tenerife two Italian hotel guests tested positive for the virus. This resulted in hundreds of guests at the H10 Costa Adeje Palace hotel being isolated to facilitate further

testing and to halt the spread of the disease. Later in the day, the first case was reported in mainland Spain, with a woman from Catalonia, who had recently returned from Italy, diagnosed in Barcelona. A seventh case was confirmed in the late evening by the Madrid regional government.

- Switzerland's government announced the first infection in the country.
- Taiwan confirmed one more case, an 11-year-old boy, bringing the total number to 31.
- Thailand reported two more cases, taking its total to 37. The two new cases, both Thai nationals, were people who were being monitored due to the travel history of others who had travelled to countries with infection risks.
- The fourth death associated with Diamond Princess was confirmed.

26 February

- The WHO Director-General, Tedros Adhanom, noted that "...14 countries that have had cases have not reported a case for more than a week, and even more importantly, 9 countries have not reported a case for more than two weeks: Belgium, Cambodia, Finland, India, Nepal, Philippines, the Russian Federation, Sri Lanka and Sweden."
- Australia confirmed one more case, bringing the total number to 23.
- Bahrain confirmed three additional cases, bringing its total to 26. Later in the evening it confirmed another 7 cases, taking the total to 33.
- Canada confirmed a new case in Toronto of a woman with travel history to Iran. This took the country's total to 12.
- Croatia confirmed its second and third cases, the second being the twin brother of the first patient and the third being a man who works in Parma, Italy.
- Finland reported its second case.
- France confirmed three new cases, taking it to 17 cases. It also announced its second death from the disease, a 60-year-old French man who had been diagnosed the night before. An eighteenth case was later declared.
- Georgia reported its first case in the country, a Georgian native travelling from Iran.
- Germany reported five additional cases. By the end of the day the overall total number of cases in the country had risen to 27.

- Greece confirmed its first case, a 38-year-old woman who had recently travelled to Italy.
- Hong Kong reported six new cases, including a 16-year-old boy and his 21-year-old sister who were both aboard Diamond Princess, bringing the total to 91 cases.
- Italy confirmed 51 additional cases, bringing its total to 374. It also confirmed the twelfth death from the virus. In the evening they declared 27 new cases followed by an additional 54 cases, with two earlier cases declared false positives.
- Iran confirmed four more deaths, taking the total to 19. Another 44 cases were confirmed, bringing the total number to 139.
- Japan confirmed its second death from the virus. It also reported 5 new confirmed cases in Nagoya. A third death was later confirmed.
- Kuwait confirmed that their total of confirmed cases had risen to 12 with one new case. Later that day they confirmed the overall total had risen to 25.
- Lebanon confirmed its second case.
- North Macedonia confirmed its first case, a woman who had recently returned home from Italy.
- Norway confirmed its first case, a person who had returned from China.
- Pakistan confirmed its first case, a 22-year-old man from Sindh province who had travelled to Iran, and also confirmed a second case with no further details.
- Romania confirmed its first case.
- Russia confirmed three more cases, who were passengers on the cruise ship Diamond Princess, bringing the total number to 5.
- Singapore confirmed two more cases, bringing the total number to 93.
- Spain confirmed five more cases including a man who was hospitalised in Seville after testing positive, bringing the total number to 12. This was the first confirmed case in the southern region of Andalusia. Another case was later confirmed in the Canary Islands.
- South Korea confirmed 169 more cases, bringing the total number to 1,146. It also recorded an additional death, increasing its total to 12. In the afternoon, the country reported 115 more infections, bringing the total number to 1,261.
- Sweden reported a new case, the second in the country.

- Taiwan confirmed an additional case, bringing its total to 32.
- Thailand confirmed three additional cases, bringing its total to 40 cases.
- The United States confirmed three new cases - one being a domestic case in California with no travel history, and the other two being former passengers aboard Diamond Princess - bringing the total number to 60.
- Diamond Princess saw 14 new confirmed cases, taking it to 705 cases aboard the ship. It is estimated that more passengers on the ship could be infected than previously thought.
- For the first time, a case was re-imported back to China from another country; the patient travelled from Iran.

27 February

- Austria reported its third case, the first in Vienna.
- Canada confirmed the first human-to-human transmission in Toronto, the husband of the woman who had travelled to Iran, bringing the total cases in the country to 13. The country later reported a presumptive case in Quebec. The patient had travel history to Iran. This presumptive case was the first case in Quebec, bringing the total number to 14.
- China reported through the National Health Commission 433 new cases and 29 new deaths across the mainland. Of those, 409 cases and 26 deaths were in Hubei province.
- Denmark confirmed its first case, a person who had recently returned from Italy.
- Estonia reported its first case, an Iranian man who had recently returned from Iran.
- France reported 20 new cases, raising the number of cases in the country to 38.
- Germany saw another case reported in North Rhine-Westphalia. In the evening it was reported that there were 14 new cases in the same state, bringing Germany's total to 40 cases. The cases rose to 45 by the end of the day.
- Greece confirmed two more cases, the first patient's daughter and another woman in Athens, bringing the total to 3 cases.

- Hong Kong reported two additional cases, both women infected after visiting a Hong Kong temple at the centre of a COVID-19 cluster. This took the total number of cases to 93.
- Iran reported that 22 people had now died amid 141 people infected by the virus. The chairman of Iran's National Security Committee, Mojtaba Zolnour, announced he had been infected. Iran later announced an increase to 26 deaths and 245 cases. The country's Vice President for Women and Family Affairs Masoumeh Ebtekar also tested positive.
- Iraq reported a new case associated with travelling to Iran, the sixth in the country and the first in Baghdad. A seventh case was confirmed in Kirkuk later in the day.
- Israel reported an additional case, a person who had travelled to Italy, bringing the total number to 3.
- Italy reported that two more people had died, bringing the death toll to 14. The Civil Protection agency reported the number of confirmed cases had risen from 420 to 528. Later in the day it was announced that three more people had died, taking the death toll to 17, and the total confirmed cases had risen to 650. At the end of the day, the total cases rose to 655. Mayor Pietro Mazzocchi of Borgonovo Val Tidone was diagnosed with the virus.
- Japan reported 17 new cases. It confirmed a case involving a recovered patient being reinfected with COVID-19. The total number of local cases had increased to 186. An eighth person died, a Japanese man in his 80s, who lived in the northern-most island of Hokkaido. Additionally, Prime Minister Shinzo Abe asked that all elementary, middle, and high schools shall be closed until late March.
- Kuwait confirmed the total cases in the country were 43, with all having recently travelled to Iran.
- Lebanon reported a third case with the person having travelled from Iran.
- Malaysia confirmed an additional case, a woman who had recently travelled to Japan, bringing the total to 23. That made it the first case to be imported from Japan.
- The Netherlands reported its first case, a person who had been skiing in Italy.

- Nigeria confirmed its first case, an Italian citizen who worked in Nigeria and returned from Milan. It was the first case of coronavirus in sub-Saharan Africa.
- Norway confirmed three additional cases, two in Oslo and one in Bærum, taking the total to four cases. Two returned from Italy and one from Iran.
- Oman reported an additional case, raising the country's total to 5. An additional case was confirmed later that day, who had travelled from Iran.
- San Marino confirmed its first case.
- Singapore confirmed three more cases, bringing the total number to 96.
- South Korea reported 334 new cases, bringing the total number to 1,595. 171 additional cases were later confirmed, taking the total number to 1,766.
- Spain saw a new case reported in Valencia and two more in Madrid. A further case was reported in Barcelona. By the early evening, the total number of confirmed cases had risen to 22. In the evening six additional cases were declared in Valencia.
- Sweden had five new cases confirmed, one in the Uppsala region, three people in Västra Götaland, and one in the Stockholm region.
- Switzerland reported more cases, with a total of six confirmed cases across five cantons. By the end of the day, a total of eight cases had been detected.
- The United Arab Emirates reported the recovery of two patients, and six new cases, bringing the total number to 19. It also saw two suspected cases attached to the UAE Tour, a cycling event, which resulted in the cancellation of the event.
- The United Kingdom confirmed two more cases, contracted in Italy and Tenerife, taking the total to 15 cases. In the evening, a sixteenth case was confirmed, and the first case in Northern Ireland, someone who had come from Italy via Dublin.

28 February

- Press release of the Swiss government banning events of more than 1,000 people (28 February 2020).
- Australia confirmed one additional case, a former Diamond Princess passenger now in Western Australia, bringing the total number to 24.

- Azerbaijan confirmed its first case, a Russian national who recently returned from Iran.
- Bahrain announced three new cases, raising the total to 36. An additional two cases were later confirmed, both having recently travelled from Iran, one being a Saudi Arabian national.
- Belarus confirmed its first case, a student from Iran.
- Canada reported two positive cases in Ontario. One of them had travel history to Iran, while the other had visited Egypt. The total number of cases rose to 16.
- China confirmed an additional 327 new confirmed cases, of which 318 were in the province of Hubei. There were 44 deaths of which 41 were in Hubei, one in Xinjiang and two in Beijing.
- Croatia confirmed two new cases, one being the girlfriend of the first patient and the other the wife of the third patient, bringing the total number to 5.
- Denmark confirmed an additional case, bringing the total number to 2.
- Finland confirmed an additional case, taking it to 3 confirmed cases.
- France announced two new cases, bringing the total number to 40. The total number of cases later rose to 57.
- Georgia confirmed a second case, a citizen who had returned from Italy. More than 18 people were reported to be in quarantine.
- Germany reported cases totalling "almost 60", according to a Health Ministry official.
- Greece confirmed an additional case, who had recently travelled to Italy. This was the country's fourth case.
- Hong Kong confirmed an additional case, a patient from Diamond Princess, bringing the total number to 94.
- Iceland confirmed its first case, a person who had returned from Italy.
- Iran confirmed 143 new cases and 6 more deaths, with a total of 388 cases and 32 deaths. Four MP's had tested positive, an increase from the two who had already announced their cases.
- Iraq confirmed an eighth case, a woman who recently came back from Iran.
- Israel confirmed an additional case, the second in the country - the wife of the previous person diagnosed. Five additional cases were later

confirmed, taking the total to 7. One of the new cases was an Israeli who had recovered in Japan, who was diagnosed again after returning.

- Italy reported that four more people had died, bringing the death toll to 21. The number of confirmed cases had risen from 650 (the day before) to 888.
- Japan announced an additional death, bringing the total to 5. They also confirmed 12 new cases in Hokkaido. Japan also later confirmed another death regarding a British native on board Diamond Princess, marking the first foreign victim on the cruise ship.
- Kuwait announced two additional cases, taking its total to 45.
- Lebanon confirmed its fourth case, a Syrian man, the first case of local transmission.
- Lithuania confirmed its first case, a woman who returned from Italy.
- Malaysia confirmed two more cases, bringing the total number to 25.
- Mexico confirmed its first two cases, one in Mexico City and one in Culiacán. Both cases were individuals who recently returned from a trip to Bergamo, Italy.
- Monaco saw the government confirm the first case in the principality.
- New Zealand confirmed its first case, an individual who had returned from Iran to Auckland earlier in the week.
- The Netherlands confirmed its second case.
- Norway confirmed two other cases, with the second having serious implications. First it was confirmed that a person from Bergen, who returned from Italy, tested positive. Later it was confirmed that a person working at Ullevaal Hospital in Oslo also tested positive, after having returned from Italy. According to standard recommendation he was told to go to work, since he had no symptoms, and he had worked with a significant number of patients for a couple of days before testing positive. Norway now has 6 confirmed cases.
- Romania confirmed two new cases, both having recently returned from Italy.
- Singapore confirmed two more cases, bringing the total number to 98.
- Spain confirmed more cases, taking it to 32 cases with 5 in Madrid, 8 in Comunidad Valenciana, 6 in the Canary Islands, 6 in Catalonia, 1 in the Balearic Islands, 6 in Andalucía, 1 in Aragón and 2 in Castilla y León.

- South Korea confirmed 256 new cases, bringing the total number of infected in the country to 2,022. 182 of the new cases were in Daegu. There were three new deaths, raising the number to 16. Later, the number of cases there shot up by 315, bringing the total number to 2,337.
- Sweden confirmed four more cases, raising the country's total to 11.
- Switzerland's cases increased to 15.
- Taiwan confirmed two additional cases, taking its total to 34.
- Thailand confirmed an additional case, taking the overall number to 41.
- The United Kingdom reported its 17th and 18th cases, in people who had travelled from Iran, and its 19th case and the first in Wales, being someone who had travelled from Northern Italy. In the evening, the 20th case was confirmed, the first case of being passed on in the UK from an unknown source.
- The United States confirmed 4 more cases, including 2 former passengers of Diamond Princess. Washington state authorities later confirmed two additional presumptive cases, bringing its total to 66. One had recently returned from South Korea, and the other case was unrelated and locally acquired.
- SARS-CoV-2 was detected for the first time in tears and conjunctival secretions (pink eye) of a patient. This was only detected in 1 out of the 30 patients that were tested. The excerpt from the study stated, "In the study of this small sample, we used conjunctival test paper to obtain tears and conjunctival secretions of 30 patients for standard RT-PCR assay. Only one patient with conjunctivitis found viral RNA in his tear fluid and conjunctival secretion twice."

29 February

- Australia confirmed one more case, bringing the total number to 25.
- Austria confirmed four new cases, bringing the total to 10.
- Azerbaijan reported two new cases, both recently returned from Iran.
- Bahrain confirmed three additional cases, all who had travelled to Iran, bringing the total number to 41.
- Brazil confirmed the second case, a 32-year-old man who arrived from Milan, Italy.
- Canada confirmed four new cases, three in Ontario and one in British Columbia, bringing the total to 20.

- China confirmed 427 new cases - 423 of which were in Hubei province - bringing the total number to 79,251. Deaths increased by 47, to a total of 2,835.
- Croatia confirmed its sixth case, a close relative of the third and fifth patients.
- Denmark confirmed its third case, a person recently returned from a trip to Munich, Germany, where he came into contact with someone who was later determined to have the virus.
- Ecuador confirmed its first case, a woman who recently returned from a trip to Spain.
- France confirmed 16 additional cases, bringing the total number to 73. Additional cases were confirmed, taking the number to 100.
- Georgia confirmed one new case, taking the total number to 3.
- Germany's confirmed cases increased to 66.
- Greece confirmed three new cases, two of which were close contacts of a previous confirmed case, the other had travelled to Italy.
- Iran confirmed 205 new cases, and 9 new deaths, bringing the total to 593 infected and 43 dead.
- Iraq confirmed 5 new cases, bringing the total number to 13. Four of them are in Baghdad and the fifth is in Babil province.
- Ireland confirmed its first case, associated with travel to northern Italy.
- Italy confirmed 239 new cases and 8 new deaths, bringing it to 1,128 cases and 29 deaths.
- Lebanon confirmed three more cases, bringing the total number of cases to seven.
- Luxembourg confirmed its first case.
- Mexico confirmed two more cases, one man in the State of Mexico who participated in the same trip to Bergamo, Italy, that the first two infected made, and a woman in Coahuila who traveled to Milan, Italy, between January and February, bringing the total to 4.
- The Netherlands confirmed four additional cases, bringing the total to 6. A seventh case was later confirmed.
- Norway confirmed nine new cases, bringing the total to 15.
- Pakistan confirmed two mores cases, bringing its total to 4. One is a person from Karachi who had recently traveled to Iran, and the other case was in the Federal Area.

- Qatar confirmed its first case, a Qatari national who had recently returned from Iran.
- Singapore confirmed four more cases, bringing the total number to 102.
- South Korea confirmed 594 more cases, bringing the total number to 2,931. This later increased with another 219 cases declared, bringing the total number to 3,150. Another death was confirmed, bringing the total to 17. The first reinfection case was also confirmed.
- Spain confirmed 26 more cases, bringing the total number to 58.
- Sweden reported two new cases, taking the total to 13.
- Switzerland confirmed 3 more cases, bringing the total number to 18.
- Taiwan confirmed five more cases, bringing the total number to 39.
- Thailand confirmed one more case, bringing the total number to 42.
- The United Kingdom confirmed three additional cases, taking the total to 23.
- The United States confirmed its first death, a man from Washington near the Seattle area. The country also reported two more cases, bringing the total number to 68.

1 March

- Armenia confirmed its first case, a 29-year-old man returning from Iran.
- Australia confirmed the first death in the country, a 78-year-old man from the cruise ship Diamond Princess. One additional case was confirmed, bringing the total number to 26.
- Austria confirmed four more cases, bringing the total number to 14.
- Bahrain confirmed six more cases, bringing the total number to 47.
- Canada confirmed four new cases in Ontario, bringing the total number to 24. All four have travel history to Iran or Egypt.
- China confirmed 573 new cases; 570 of which were in Hubei province, bringing the total number to 79,824. Deaths increased by 35 to a total of 2,870.
- Czech Republic confirmed the first three cases in the country, individuals that have travelled from Northern Italy.
- Dominican Republic confirmed its first case, a 62-year-old Italian tourist.
- Ecuador confirmed five more cases, bringing the total to six.

- Egypt confirmed one more case, bringing the total number to 2.
- Finland confirmed two more cases related to woman diagnosed on 27 February, bringing the total number to 6.
- France confirmed 30 more cases, bringing the total number to 130.
- Germany confirmed 63 more cases, bringing the total number to 129.
- Hong Kong confirmed three more cases, bringing the total number to 98.
- Iran confirmed 385 new cases with 11 more deaths, bringing the total number to 978 and 54 respectively. A total of 23 members of Parliament of Iran, about 8%, have tested positive for the virus.
- Iraq confirmed six more cases, bringing the total number to 19.
- Israel confirmed three more cases, bringing the total number to 10.
- Italy confirmed 566 new cases and five deaths, bringing the total number to 1,694 and 34 respectively.
- Japan confirmed another death from the coronavirus, bringing the total to 6.
- Malaysia confirmed four more cases, bringing the total number to 29.
- Mexico confirmed one additional case, bringing the total number to 5.
- The Netherlands confirmed three more cases, bringing the total number to 10.
- Saint Barthélemy confirmed its first case.
- Saint Martin confirmed its first case.
- San Marino confirmed seven more cases, bringing the total number to 8. The first death was also confirmed.
- Scotland confirmed its first case.
- Singapore confirmed four more cases, bringing the total number to 106.
- South Korea confirmed 376 more cases, bringing the total number to 3,526. Another 210 cases were later confirmed with the 18th death reported, bringing the total number to 3,736.
- Spain confirmed 26 more cases, bringing the total number to 84. Spain also confirmed its first death due to the outbreak, in Valencia.
- Sweden confirmed one more case, bringing the total number to 14.
- Thailand confirmed its first death from the coronavirus, a 35-year-old retail worker who also had dengue fever.
- The United Kingdom confirmed an additional thirteen cases of the virus, including one person from Essex who had not travelled to any

country previously affected by the virus, bringing the total in the country to 35.

- The United States reported the second confirmed death in Washington state. The first cases in Rhode Island, Florida, and New York were confirmed. The authorities confirmed 21 more cases in total, bringing the number to 89.

2 March

- Andorra confirmed its first case.
- Australia confirmed four new cases, bringing the total number to 30, including its first human-to-human transmission.
- Bahrain confirmed two more cases, bringing the total number to 49.
- Canada confirmed three more cases, all in Ontario, bringing the total number to 27.
- China confirmed 202 new cases, bringing the total number to 80,026. Deaths increased by 42 to a total of 2,912.
- France confirmed 61 more cases, bringing the total number to 191. The third death was also reported.
- Germany confirmed 21 more cases, bringing the total number to 150.
- India reported its first cases since the initial outbreak. One case was detected in New Delhi, while the other is from Telangana. Another confirmed case was later detected in Jaipur, bringing the total number to 6.
- Indonesian president Joko Widodo announced the first two confirmed cases in the country. The two people contracted the virus from a Japanese in Depok, who later tested positive in Malaysia. The mother and daughter are now hospitalized in North Jakarta.
- Iran's total cases rose to 1,501 with 66 deaths.
- Iraq confirmed two more cases, bringing the total number to 21.
- Israel confirmed two more cases, bringing the total number to 12.
- Italy confirmed 342 new cases and 18 deaths, bringing the total number to 2,064 and 52, respectively.
- Jordan confirmed its first case.
- Kuwait confirmed ten more cases, bringing the total number to 56.
- Latvia confirmed its first case, a woman who had flown from Milan to Riga through Munich.
- Morocco saw its first case confirmed.

- The Netherlands confirmed eight new cases, bringing the total number to 18.
- Portugal confirmed its first two cases, one of whom had returned from Italy, the other from Spain.
- Qatar confirmed four new cases, taking their total to 7.
- Russia confirmed one additional case, bringing the total to 6 cases.
- Saudi Arabia confirmed their first case in a citizen who had travelled to Iran and returned to Saudi Arabia via Bahrain.
- Senegal confirmed its first case, where the person had recently travelled from France.
- Singapore confirmed two more cases, bringing the total number to 108.
- South Korea confirmed 476 more cases, bringing the total number to 4,212. Four more deaths were confirmed, bringing the total to 22. Another 123 cases were later confirmed, bringing the total number to 4,335.
- Spain confirmed 36 more cases, bringing the total number to 120.
- Sweden confirmed one more case, bringing the total number to 15.
- Taiwan confirmed one additional case, bringing the total to 41 cases.
- Thailand confirmed one more case, bringing the total number to 43.
- Tunisia confirmed its first case.
- The United Kingdom confirmed three additional cases, taking the total to 39. An original fourth diagnosis of positive of a person was confirmed later as negative.
- The United States confirmed 13 more cases, bringing the total number to 102. Five more deaths were also confirmed, bringing the total number to 6.

3 March

- Argentina confirmed its first case, a person who had recently returned from Italy.
- Australia confirmed eight more cases, bringing the total number to 38.
- Canada confirmed six more cases, two in Ontario and four in British Columbia, bringing the total to 33.
- Chile confirmed its first case.

- China confirmed 125 new cases, the lowest number of new cases since January, bringing the total number to 80,151. 31 new deaths were also confirmed, bringing the total to 2,943.
- Ecuador confirmed three more cases, bringing the total number to 10.
- France confirmed 21 more cases, bringing the total number to 212. The fourth death was also reported.
- Germany confirmed 38 more cases, bringing the total number to 188.
- Gibraltar confirmed its first case, a person who had travelled back from Northern Italy via Malaga airport.
- Iran confirmed that 23 MPs were diagnosed with the virus. The country confirmed 835 more cases with 11 more deaths, bringing the total number to 2,336 and 77, respectively.
- Ireland confirmed one more case, bringing the total number to 2.
- Italy confirmed 466 new cases and 28 deaths, bringing the total number to 2,502 and 80, respectively.
- Liechtenstein confirmed its first case.
- Malaysia confirmed seven more cases, bringing the total number to 36.
- San Marino confirmed two new cases, bringing the total number to 10.
- Singapore confirmed two more cases, bringing the total number to 110.
- South Korea confirmed 600 more cases, bringing the total number to 4,812. 12 more deaths were also confirmed, bringing the total to 34. 377 more cases were confirmed in the afternoon, along with two more deaths, taking total to 5,186 cases and 36 deaths.
- Spain confirmed the first death in the country, a person who died on 13 February, making him the earliest recorded death in Europe. The country reported 31 more cases, bringing the total number to 151.
- Sweden confirmed 15 more cases, bringing the total number to 30.
- Ukraine confirmed its first case, a person having travelled from Italy via Romania.
- The United Kingdom confirmed 12 new cases, bringing the total number to 51.
- The United States confirmed three more deaths, bringing the total to 9. The first case of coronavirus in North Carolina was reported, coming from the nursing home in Washington state responsible for the first deaths from the virus. Another case in Florida was presumptively

confirmed positive, bringing the total number of Floridian cases to three. The number of cases stood at 126, increasing by 24 from 102.

4 March

- Brazil confirmed two more cases, bringing the total number to 4.
- Canada confirmed one more case in British Columbia, bringing the total number to 34.
- China confirmed 119 new cases, bringing the total number to 80,270. 38 new deaths were also confirmed, bringing the total to 2,981.
- Faroe Islands confirmed its first case.
- France confirmed 73 more cases, bringing the total number to 285.
- Germany confirmed 52 more cases, bringing the total number to 240.
- Hong Kong confirmed a case of human-to-animal transmission involving a pet dog.
- Hungary confirmed its first cases, two Iranian students who are asymptomatic.
- India confirmed 23 more cases, including 15 Italian tourists, bringing the total number to 29.
- Iraq confirmed its first death from the coronavirus.
- Italy confirmed 28 more deaths, bringing the total to 107. This made it the highest jump in a single day. The country confirmed 587 new cases, bringing the total number to 3,089.
- Cases in Japan have topped the 1,000 mark (including 706 cases on the cruise ship Diamond Princess which the World Health Organization classifies as being located "on an international conveyance" and not in Japan) with the first confirmed case in Yamaguchi prefecture.
- Malaysia confirmed 14 more cases, bringing the total number to 50.
- New Zealand confirmed its second case of the coronavirus. The infected individual had returned the previous week from Italy with her family on an Air New Zealand flight from Singapore to Auckland.
- Poland confirmed its first case.
- San Marino confirmed five more cases, bringing the total number to 15.
- Singapore confirmed two more cases, bringing the total number to 112.
- Slovenia confirmed its first case. A person travelled through Italy.

- South Korea confirmed 516 more cases, bringing the total number to 5,328. Another 293 cases were confirmed in the afternoon, bringing the total number to 5,621.
- Sweden confirmed 22 more cases, bringing the total number to 52.
- The United Kingdom confirmed 34 new cases, then Northern Ireland confirmed an additional two, bringing the total number to 87.
- The United States confirmed 11 total deaths, with the first death outside of Washington state in California. California's total of infections has jumped to 51. The country confirmed 33 new cases, bringing the total number to 159.

5 March

- Algeria confirmed nine additional cases, bringing the total number to 17.
- Belgium confirmed 27 new cases, bringing the total number to 50.
- Brazil confirmed three new cases, bringing the total to 8.
- Bosnia and Herzegovina confirmed the first two cases in the country.
- Canada confirmed 11 new cases in Ontario, and the first one in Alberta, bringing the total to 45.
- China confirmed 139 new cases, bringing the total number to 80,409. 31 new deaths were also confirmed (all in Hubei), bringing the total to 3,012.
- Ecuador confirmed three more cases, bringing the total number to 13.
- Egypt confirmed the first case in an Egyptian national, who had recently travelled from Serbia via France.
- France registered 92 new confirmed cases, taking the total number of cases to 377. It also saw deaths rise from 4 to 6. This then rose to 423 total cases and 7 deaths. Jean-Luc Reitzer, was the first Assembly member to be diagnosed with the virus.
- Georgia confirmed five new cases, all of whom had recently returned from Italy.
- Germany confirmed 87 new cases, taking it to a total of 349.
- Greece confirmed its tenth case. It later confirmed an additional 21 cases, taking it to a total of 31.
- India confirmed one more case, bringing the total number to 30.
- Iran saw its largest daily increase in cases, with 591 new cases, taking it to a total of 3,513 confirmed cases. It also saw 15 new deaths,

increasing its total to 107. Hossein Sheikholeslam, a diplomat and a former member of parliament and Iran's former ambassador to Syria died from the virus.

- Ireland confirmed six new cases, bringing the total number to 13.
- Israel confirmed two new cases, bringing the total to 17. And of these is a bus driver from East Jerusalem who drove a group of Greek tourists in Israel and the West Bank, the other having recently returned from Italy.
- Italy reported 769 new cases and 41 new deaths, bringing the total number to 3,850 and 148 respectively.
- Martinique confirmed its first two cases.
- The Netherlands confirmed 44 new cases, bringing the total number to 82.
- New Zealand confirmed its third case, an Auckland man who had contracted the virus after family members had returned from a trip to Iran. Relatives of the man at two Auckland schools, Auckland Grammar School and Ormiston Junior College, had also been placed into isolation.
- Norway confirmed six new cases, bringing the total number to 66.
- Pakistan confirmed a sixth case, a 69-year-old man in Karachi.
- Palestine reported its first cases in the West Bank city of Bethlehem. Seven staff at a hotel were reported to be infected by visiting tourists from Greece. These tourists were the same ones who the Israeli bus driver was infected from.
- Russia confirmed one more case, bringing the total number to 7.
- Saudi Arabia confirmed three additional cases, bringing the total number to 5.
- Scotland confirmed three more cases, bringing the total to 6.
- Singapore confirmed five more cases, bringing the total number to 117.
- South Africa records its first case in the KwaZulu-Natal province.
- South Korea confirmed 145 more cases, bringing the total number to 5,766. The death toll had reached 35.
- Spain have confirmed that the total amount of cases had risen to 234.
- Sweden confirmed 42 more cases, bringing the total number to 94.
- Switzerland reported its first death from the virus.

- The United Kingdom's total increased to 90. The UK also later confirmed the total number of cases had increased further up to 116, as well as recording the first death in the country, an older person with underlying health conditions.
- One additional death in Washington State brought the U.S. total to 12. The United States saw a significant increase in cases, with 31 new cases in Washington state. In New York state, cases doubled to 22. Colorado's first two cases were confirmed. In total, the US confirmed 69 more cases, bringing the total number to 228. Three more deaths were reported, bringing the total number to 14.

6 March

- Austria announced 13 new cases, taking its total to 55.
- Belgium confirmed 59 new cases, bringing the total to 109.
- Bhutan announced its first case, an American tourist who had recently also travelled to India after leaving the United States on 18 February.
- Brazil confirmed five new cases, bringing the total number to 13.
- Cameroon confirmed its first case, a French citizen.
- Canada confirmed four new cases in Ontario, bringing the total to 51 with 26 from Ontario. A husband and wife returned from the cruise ship Grand Princess in California and two men returned from Iran and Las Vegas.
- China confirmed 143 new cases, bringing the total number to 80,552. 30 new deaths were also confirmed, bringing the total to 3,042. Hubei had no cases as a result. In addition, the country detected 4 new cases in Beijing, all visiting Italian nationals.
- Colombia confirmed its first case, a woman who recently travelled from Italy.
- Costa Rica has confirmed its first case involving an American from New York.
- Egypt confirmed 12 additional cases, bringing the total number to 15.
- France confirmed 190 new cases and two additional deaths, bringing the total to 613 cases and 9 deaths.
- Germany confirmed 185 new cases, taking it to a total of 534. 105 more cases were later confirmed, bringing the total number to 639.
- The cruise ship Grand Princess announced it had 21 people on board testing positive.

- Greece confirmed 14 additional cases.
- Iceland confirmed six additional cases, bringing the total to 43.
- India confirmed one more case, bringing the total number to 31.
- Indonesia confirmed two more cases, bringing the total number to four.
- Iran confirmed 1,234 new cases and 16 new deaths, taking its totals up to 4,747 cases and 124 deaths. They also reported 913 people had recovered.
- Italy announced 778 new cases and 49 new deaths.
- Malaysia confirmed 28 new cases, the largest daily increase in the country's number of confirmed cases. This took the country's total to 83.
- The Netherlands reported its first death, and 44 new cases, bringing it to a total of 128 cases.
- New Zealand confirmed its fourth case: a man who is the partner of the second case confirmed on 3 March.
- North Macedonia confirmed two additional cases, taking its total to 3.
- Peru confirmed its first case.
- Poland confirmed four additional cases (two people that travelled from Italy, one person who travelled from United Kingdom and one person who travelled from Germany with the first case confirmed day earlier), taking their total to 5.
- Portugal confirmed four additional cases (three in Porto and one in Lisbon), bringing the total to 13. All three cases in Porto have a connection with Italy.
- Qatar confirmed three new cases, taking it to a total of 11.
- Russia confirmed six more cases, bringing the total number to 13.
- Serbia confirmed its first case.
- Singapore confirmed 13 new cases, bringing the total number to 130. This is the largest increase of cases in a single day so far.
- Slovakia confirmed its first case.
- Spain confirmed 104 additional cases, bringing the total number to 360.
- Sweden confirmed 43 additional cases, bringing the total number to 137.
- Switzerland announced 123 additional cases, raising it to a total of 210 cases and one fatality.

- Togo confirmed its first case.
- In the United Kingdom, the total number of cases were confirmed as 163, a rise of 47 and the highest day-on-day increase.
- In the United States, the first three cases were confirmed in the state of Maryland. Two deaths in Florida were confirmed. Officials in the US state of Hawaii have confirmed their first case, a former passenger of the cruise ship Grand Princess. In total, the US confirmed 104 more cases, bringing the total number to 332. One more death in Washington State plus two in Florida brought the total number of U.S. deaths to 17. WHO has different data: 19 new cases, 148 in total. 1 new death, 10 in total.
- The Vatican City confirmed its first case.
- Vietnam confirmed one more case, bringing the total number to 17. It was the first case in three weeks.

7 March

- Afghanistan confirmed three additional cases, taking the total to 4.
- Argentina confirmed its first death, also the first in South America, a 64-year-old man who had travelled to Paris.
- Bahrain confirmed 19 new cases, taking the total number to 79.
- Belgium confirmed 60 additional cases, bringing the total number to 169.
- Brazil confirmed six new cases of COVID-19, raising the total number to 19.
- Canada confirmed six additional cases, taking the total to 57.
- Egypt confirmed 33 new cases on a Nile cruise ship.
- France's total rose to 949 cases and 16 deaths.
- Germany confirmed 45 new cases, taking it to a total of 684. Another 116 cases were later confirmed, bringing the total number to 800.
- Iran confirmed 1,076 new cases and 21 additional deaths, taking it to a total of 5,823 infected and 145 dead. They also confirmed 16,000 cases were hospitalised as suspect cases and 1,669 cases had recovered. Another MP, Fatemeh Rahbar, died. She had recently been elected to the parliament.
- Iraq confirmed 8 new cases taking the total to 46.
- Italy confirmed 1,247 new cases and 36 new deaths.
- Maldives confirmed its first cases, two overseas hotel employees.

- Malaysia confirmed 10 more new cases, bringing the total number to 93.
- Malta reported its first three cases, an Italian family who are residents in Malta. They had gone on holiday to northern Italy and had been in self-quarantine before being tested for the coronavirus. They are now in isolation at Mater Dei Hospital.
- Moldova confirmed its first case, a person taken to hospital after arriving on a flight from Italy.
- New Zealand confirmed its fifth case of the coronavirus: a woman who was the partner of the third COVID-19 case confirmed in New Zealand.
- Paraguay confirmed its first case.
- Peru confirmed five more cases, bringing the total to six.
- Poland confirmed an additional case, bringing the total to 6.
- Singapore confirmed eight more cases, bringing the total number to 138.
- South Korea confirmed 274 new cases and 4 new deaths, taking total cases to 7,041.
- Spain confirmed 70 new cases and 2 deaths. The total number of cases was now 516 cases.
- The United Arab Emirates confirmed 15 additional cases, taking its total to 45.
- In the United Kingdom, 42 additional cases and one additional death were confirmed, taking the total to 206 cases and 2 deaths. 3 additional cases were confirmed in Northern Ireland later in the day, taking the total to 209 cases.
- In the United States, the number of deaths rose to 19 with 16 in Washington, 1 in California, and 2 in Florida. The total for the country increased to 444 confirmed cases. WHO has different data: 65 new confirmed cases, totalling 213. And 1 new death, 11 in total.
- Vietnam confirmed three new cases, bringing the total number to 20.

8 March

- Albania confirmed the first two cases in the country.
- Austria confirmed 25 additional cases, taking its total to 104.
- Bahrain confirmed 17 new cases, taking the total to 94.

- Bangladesh confirmed its first three cases, two people who had come from Italy separately, and a contact of one of the cases.
- Belgium confirmed 31 additional cases, bringing the total number to 200.
- Brazil confirmed six new cases of COVID-19, raising the total number in the country to 25.
- Bulgaria confirmed its first four cases.
- Canada confirmed 12 new cases, bringing the total number to 69.
- Egypt confirmed its first death (and the first death in Africa) - a German national who was hospitalised on 1 March and then suffered respiratory failure caused by acute pneumonia on 7 March.
- France confirmed that the number of cases had risen to 1,126 and 19 deaths. Elisabeth Toutut-Picard became the second Assembly member to be diagnosed with the virus.
- Germany's cases rose to over 1,000 with 1,018 cases confirmed.
- Hong Kong confirmed four new cases, taking the total to 114, and a third death.
- India confirmed five additional cases, all in Kerala. This took the country's total up to 39.
- Indonesia confirmed two additional cases, taking them to 6 cases. One of the new cases is a male crew member from the Diamond Princess.
- Iran confirmed 49 additional deaths, the highest daily toll in the country, taking it to a total of 194 deaths. Total cases have risen to 6,566, an increase of 743.
- Israel confirmed 11 additional cases, taking its total to 29.
- Italy confirmed an additional 1,492 cases and 133 deaths, taking it to a total of 7,375 cases and 366 deaths.
- Kuwait confirmed an additional case, taking its total to 62.
- Latvia confirmed an additional case.
- Lebanon confirmed four more cases, bringing the total number to 32.
- Malaysia confirmed an additional six cases, taking the total to 99.
- Poland confirmed two additional cases, taking the total to 8. In the evening, three more cases were confirmed, bringing the total number to 11.
- Portugal confirmed four additional cases.
- Qatar confirmed three additional cases, taking their total to 15.

- Saudi Arabia confirmed four additional cases, bringing the total number to 11.
- Singapore confirmed 12 new cases, bringing the total number to 150.
- Spain announced the new total was 616 cases, along with 17 deaths, an increase of seven.
- South Africa confirmed its third case, the wife of the first diagnosed case.
- South Korea confirmed an additional 93 new cases, taking its total to 7,134 and two additional deaths.
- Switzerland confirmed there are now 332 confirmed cases, with all those people placed in isolation. There have been two deaths.
- The United Kingdom announced an increase of 64 new cases, taking it to a total of 273 cases. The UK also confirmed the third death in the country. Five new cases were additionally confirmed in Northern Ireland.
- The United States confirmed 120 more cases, bringing the total number to 564. Two more deaths were confirmed, bringing the total number to 21. According to WHO, the number of confirmed cases was 213, total deaths 11. No new cases, no new deaths were recorded.
- Vietnam confirmed ten more cases, bringing the total number to 30.

9 March

- Australia's number of confirmed cases rose to 80, with six additional cases.
- Bahrain confirmed an additional 24 cases, taking the total to 118.
- Belgium confirmed 39 additional cases, bringing the total number to 239.
- Brazil confirmed five new cases in the state of Rio de Janeiro, raising the total number to 30.
- Brunei's health ministry confirmed a first case of a local man who returned from Kuala Lumpur on 3 March. Symptoms began on 7 March and preliminary tests indicate the person is positive.
- Canada recorded its first COVID-19-related death. The country also confirmed ten more cases, bringing the total number to 79.
- Colombia confirmed two additional cases.
- Cyprus confirmed its first two cases.
- Finland confirmed seven new cases, bringing the total number to 30.

- France saw Guillaume Vuilletet and Sylvie Tolmont confirmed as infected deputies, before Michèle Victory became the fifth deputy of the Assembly to be diagnosed with the virus. The number of cases increased to 1,412. It was also confirmed that the culture minister, Franck Riester, had tested positive.
- Germany reported its first deaths with two fatalities. Cases increased to 1,176.
- Guernsey confirmed its first case.
- Hungary confirmed two new cases.
- India confirmed five additional cases, taking the total to 44.
- Indonesia confirmed 13 additional cases, bringing the total number to 19.
- Iran announced an update to the total cases to 7,161, with a revised number of deaths to 237. New infections are 595 new cases and 43 additional deaths. Those recovered has reached 2,394.
- Ireland confirmed three additional cases, taking the country's total to 24.
- Italy saw confirmed cases rise to 9,172 and deaths to 463.
- Kuwait announced three additional cases, taking it to a total of 65.
- Lebanon reported nine new cases, taking the total to 41.
- Malaysia confirmed 18 new cases, taking it to 117.
- The Netherlands announced 56 additional cases.
- Nigeria confirmed its second case, a Nigerian who had contact with the Italian first case.
- Norway declared 11 additional cases. By the end of the day 227 cases had been confirmed.
- Oman reported two additional cases, both of whom had travelled to Iran. This took the country's total to 18, 17 of whom had travelled to Iran and one to Italy.
- Panama announced its first COVID-19 case, a 40 year old Panamanian woman from Spain.
- The Philippines confirmed 10 additional cases, doubling the current country count to 20. Another four cases were later confirmed, bringing the total number to 24.
- Pakistan saw nine new cases, bringing the total number to 16.
- Poland confirmed five more cases, bringing the total number to 16.

- Saudi Arabia confirmed four additional cases, increasing the total to 15.
- Scotland confirmed five additional case, bringing the total to 11.
- Singapore confirmed 10 more cases, bringing the total number to 160.
- South Africa confirmed another four cases, bringing the total number to 7.
- South Korea confirmed an additional 96 cases, taking it to a total of 7,478.
- Spain's numbers increased to 1,231 cases with 30 deaths, with 32 people having recovered.
- Sweden confirmed 45 additional cases, taking its total to 203. This later rose to 252 cases including the first confirmed local transmission case.
- Switzerland confirmed 42 new cases, rising to a total of 374.
- The United Arab Emirates reported 14 additional cases, taking its total to 59. Four were Emiratis and the remainder from various other nationalities.
- The United Kingdom's confirmed cases increased by two, with the confirmation of cases in Wales. Figures released in the afternoon showed the UK had risen to 317 cases. A fourth death was reported, and then a fifth.
- The United States confirmed 153 additional cases, bringing the total number to 717. Five more deaths were reported, bringing the total number to 26. According to WHO, the number of confirmed cases is 213, total deaths 11. No new cases, no new deaths were recorded.
- Vietnam confirmed an additional case, taking its total to 31.

10 March

- Australia reported 14 new cases, taking the country to a total of 107.
- Austria confirmed 25 additional cases, taking the total to 182.
- Bahrain confirmed one additional case, taking the total to 119.
- Belgium confirmed 28 new cases, bringing the total number to 267.
- Bolivia confirmed its first cases, two women who had been in Italy, arriving the country without showing any symptoms.
- Brazil confirmed four new cases, including the first case in Rio Grande do Sul, raising the total number to 34.
- Brunei confirmed six cases.

- Bulgaria confirmed two additional cases, bringing the total to 6.
- Burkina Faso confirmed its first cases with two infections, a couple who had returned from France in February.
- China saw only 19 new infections, 17 in Wuhan and 2 imported from overseas. 17 deaths were reported in Hubei.
- Democratic Republic of Congo reported its first case, a foreigner who flew into Kinshasa from Belgium who tested positive on arrival and was isolated. It was later confirmed by the Ministry of Health that the information they had originally released was incorrect and that that the first case was rather a Congolese citizen returning From France, who had contacted the Health Services two days after his arrival in Congo and had been quarantined in a local neighbourhood.
- Denmark confirmed 174 additional cases, taking it to 264 cases (including two in the Faroe Islands).
- France saw cases rise to 1,784 people and 33 people dead, with 86 people in a serious condition.
- Georgia confirmed eight additional cases, bringing the total to 23.
- Germany's number of case rose to 1,565, with the death roll remaining at 2. 341 new cases were reported in the previous 24 hours.
- Greece confirmed five new cases, bringing the total number to 89.
- Hong Kong confirmed an additional five cases, taking the total to 120 confirmed and 1 assumed case.
- India saw 18 additional cases, many in Kerala, taking the total to 62.
- Indonesia confirmed eight more cases, bringing the total number to 27.
- Iran's confirmed cases rose to 8,042 and deaths rose to 291.
- Ireland reported 10 additional cases, bringing the total number to 34.
- Israel confirmed seven more cases, taking the total to 58.
- Italy saw an increase of cases to 10,149, with deaths rising to 631.
- Jamaica confirmed its first case, a Jamaican national who travelled to the country from the United Kingdom.
- Japan confirmed 59 new cases and three new deaths.
- Jersey confirmed its first case, a person who returned from Italy.
- Kuwait announced four additional cases, increasing the total number to 69.
- Lebanon confirmed its first death from the virus.
- Malaysia confirmed 12 additional cases, taking the total to 129.

- Moldova confirmed two additional cases, taking the total to 3. The male and female, have both visited Italy in late February.
- Mongolia announced its first COVID-19 case.
- Morocco confirmed its first death, and two additional cases.
- The Netherlands confirmed 61 additional cases and a fourth death.
- Northern Cyprus reported its first case, a 65-year-old German woman who was visiting as a tourist.
- Norway confirmed 147 new cases, with total cases now standing at 374.
- Panama announced its first death and seven new cases, bringing the total number to 8.
- Pakistan confirmed an additional case, bringing the total to 20.
- Palestine recorded an additional case, taking the total to 26.
- The Philippines confirmed nine more cases, bringing the total number to 33. This comes after the tally was erroneously reported as 35.
- Poland confirmed eight additional cases, taking the total to 22.
- Qatar confirmed six additional cases, taking it to 24 confirmed cases.
- Romania confirmed eight additional cases, bringing the total number to 25.
- Russia confirmed four additional cases, bringing the total to 17.
- San Marino confirmed 11 new cases, taking it to a total of 62 cases and two deaths.
- Saudi Arabia confirmed ten additional cases, taking the total to 20.
- Scotland confirmed 7 new cases, bringing the total to 18.
- Serbia confirmed three additional cases, taking the total to 5.
- Singapore confirmed six more cases, bringing the total number to 166.
- Slovenia confirmed 15 additional cases.
- Spain confirmed 415 new cases and five new deaths. The total number of cases rose to 1,674.
- Sweden reported 78 new cases, taking the country's total to 326.
- Tunisia announced an additional three cases, taking the total to 5.
- Turkey confirmed its first case.
- The United Arab Emirates confirmed 15 additional cases, taking the total to 74.
- The United Kingdom confirmed 54 additional cases, taking it to 373 cases with the sixth confirmed death. Later this increased to 382 cases

with the nine additional cases in Wales. Nadine Dorries, Health Minister, was diagnosed with the virus.

- The United States confirmed 283 additional cases, bringing the total number to 1,000. Five more deaths were reported, bringing the total number to 31.
- Vietnam confirmed three new cases, bringing the total number to 34.

11 March

- Albania reported three new cases, taking the country's total to 15. The first death in the country was also reported.
- Armenia confirmed three new cases, bringing the total number to 4.
- Australia reported 19 new cases, taking the country to a total of 126. American actor Tom Hanks and wife Rita Wilson were diagnosed with the illness during the filming of the Baz Luhrmann's upcoming-biographical film Elvis in Australia.
- Austria confirmed 24 additional cases, taking the total to 206.
- Bahrain confirmed 77 additional cases, citizens evacuated from Iran on a flight. This brought the country's total to 189.
- Belgium reported its first death.
- Bosnia and Herzegovina confirmed seven additional cases.
- Brazil confirmed 34 additional cases, raising the total number to 69.
- Bulgaria reported its first death. It also confirmed one additional case, bringing the total to 7.
- China confirmed 11 new cases across the Mainland, in addition to 13 new cases and 22 new deaths in Hubei province.
- Colombia confirmed three additional cases, bringing the total to 6.
- Later on the same day, three more cases were confirmed, bringing the total to 9.
- Cuba confirmed its first three cases.
- Denmark confirmed 180 additional cases, raising the total to 442.
- Egypt confirmed seven additional cases, six of whom were Egyptian. The country now has 67 cases of which eight have recovered.
- France reported 487 new cases and 15 fatalities, for a total of 2,281 cases and 48 deaths.
- French Polynesia reported its first case. The person is Maina Sage, a member of the French National Assembly.
- Georgia reported five additional cases, increasing the total to 23.

- Germany saw its cases rise to 1,908.
- Greece confirmed 10 additional cases, taking the country's total to 99, 95 of whom are Greek and 4 are foreign nationals.
- Honduras confirmed its first two cases, one from Spain and one from Switzerland.
- Indonesia confirmed seven more cases, bringing the total number to 34. Its first death was also reported, confirmed by the United Kingdom's Foreign and Commonwealth Office as a 53-year-old female British national.
- Iran announced that there are now 9,000 infected, with 354 deaths. New infections were 958 with 63 additional deaths. Those recovered total 2,959. Iran's first vice president, Eshaq Jahangiri was reported infected.
- Ireland confirmed nine additional cases, bringing the total number to 43. The first death was also confirmed.
- Israel confirmed 97 cases.
- Italy announced 2,313 new cases and 196 new deaths. Total cases stood at 12,462, with deaths at 827. It was also confirmed Juventus and Italian footballer Daniele Rugani had tested positive for coronavirus.
- Ivory Coast declared its first case, a local citizen who had recently been in Italy.
- Kuwait reported three additional cases, bringing its total to 72.
- Lebanon recorded eight new cases, taking the total to 59, and a second death.
- Lithuania confirmed two additional cases, a married couple that came back from Italy, taking the total to 3 cases.
- Malaysia confirmed 20 additional cases, taking the total to 149.
- Maldives confirmed two additional cases.
- Morocco confirmed two additional cases, bringing the total to 5.
- The Netherlands' total increased to 503 cases, and one new death was reported.
- Norway confirmed 200 additional cases, taking the total number to 602.
- Paraguay confirmed three new cases.
- Panama confirmed several new cases, bringing the total to 14.

- The Philippines confirmed 16 additional cases, bringing the country's total to 49. The second death was also confirmed.
- Portugal announced 18 new cases.
- Qatar reported six new cases, increasing the country's total to 24. It later announced 238 new cases in expatriates who came in contact with three infected people, and had been quarantined. The revised number of cases is now 262.
- Réunion confirmed its first case.
- Romania confirmed 18 new cases, bringing the total to 47.
- Saudi Arabia reported an additional case, an Egyptian, taking the country's total to 21.
- Scotland confirmed 18 new cases, bringing the total number to 36.
- Singapore confirmed 12 new cases, bringing the total number to 178.
- South Africa confirmed six more cases, taking the total to 13.
- South Korea announced 242 new cases and 6 new deaths.
- Spain's cases rose to 2,231, with 54 confirmed deaths.
- Sri Lanka announced one additional case, bringing the total number to 2.
- Sweden reported 98 additional cases, taking the total to 500, and announced the first death in the country.
- Taiwan confirmed one additional case, raising the country's total to 48.
- Thailand confirmed six additional cases, taking the country's cases to 59.
- The United Kingdom confirmed an additional 86 cases, with its overall total now at 456. Two more deaths were confirmed, bringing the total to 8. The number later increased to 460, with four cases declared in Wales, including the first instance of community transmission in Wales.
- The United States confirmed 272 additional cases, bringing the total number to 1,272. Seven more deaths were reported, bringing the total number to 38. The Utah Jazz's Rudy Gobert and Donovan Mitchell were diagnosed with the illness. As a result, the NBA suspended the entire season after the night's games. The Utah Jazz vs Oklahoma City Thunder game was postponed after doctors reported Gobert had the illness.

- Vietnam confirmed three additional cases, raising the country's total to 38.
- The World Health Organisation declared COVID-19 a pandemic. The declaration followed extensive criticism that the WHO response had been weak and inappropriately favorable toward the government of China.

12 March

- Algeria confirmed five additional cases and the first death.
- Brazil confirmed 82 new cases, raising the total number to 149.
- Brunei reported five new cases, bringing the total number to 11.
- Bulgaria reported nine new cases from the capital, bringing the total number to 16. They later reported seven more cases, bringing the total number to 23.
- Colombia confirmed four more cases, raising the total to 13.
- Egypt confirmed 13 new cases and one new death.
- France's total increased to 2,876 and 61 deaths.
- Germany announced 546 new cases and two additional deaths, taking the totals to 2,745 cases and 6 deaths.
- India confirmed its first death, a 76 year old Indian national with existing health conditions, who had recently returned from Saudi Arabia.
- Italy confirmed 189 more deaths, bringing the total number to 1,016, and 2,651 new cases were also confirmed, bringing the total number to 15,113.
- Ireland confirmed 27 new cases, bringing the total to 70. 22 of the new cases were by local transmission.
- Kuwait confirmed eight more cases, taking the total number to 80.
- Malaysia confirmed its first sporadic case after 600 tests. Nine new cases were later confirmed, bringing the total number to 158.
- Norway reported its first death.
- The Philippines confirmed three more cases, bringing the total cases to 52.
- Romania confirmed one new case, bringing to total number to 48.
- San Marino repored 15 new cases and three additional deaths.
- Singapore confirmed nine more cases, bringing the total number to 187.

- St. Vincent & the Grenadines confirmed its first case.
- Spain confirmed 782 new cases and 31 new deaths, raising the country's total to 3,059 cases and 86 dead.
- Sri Lanka confirmed an additional case, taking the total number to 3.
- Switzerland confirmed 212 new cases and 2 additional deaths.
- Trinidad & Tobago confirmed its first case, a Swiss national.
- Ukraine reported two new cases, bringing the total number to 3.
- The United Kingdom confirmed 136 new cases, bringing the total number to 596. Two more deaths were also confirmed, bringing the total to 10. It was confirmed Arsenal head coach Mikel Arteta had tested positive for COVID-19.
- The United States confirmed 373 additional cases, bringing the total number to 1,645. Three more deaths were reported, bringing the total number to 41. The Utah Jazz's Donovan Mitchell tested positive for the virus.
- Vietnam confirmed six new cases, bringing the total number to 44.
- A team member of McLaren tested positive for the coronavirus, throwing the 2020 Australian Grand Prix into disarray.

13 March

- Antigua and Barbuda confirmed its first case.
- Aruba confirmed its first two cases, people who travelled in from New York.
- Bulgaria confirmed three new cases, bringing the total number to 26. Bulgaria later confirmed five additional cases, bringing the total to 31.
- Brazil confirmed 22 new cases, bringing the total number to 171. President Jair Bolsonaro tested negative for COVID-19.
- Brunei confirmed 14 new cases, bringing the total number to 25.
- The Cayman Islands confirmed its first case.
- Colombia reported three more cases, raising the total to 16.
- Costa Rica reported three new cases since 12 March, bringing the total number of cases to 26.
- Curaçao reported its first case, a Dutch tourist.
- Ethiopia confirmed its first case.
- France's numbers increased to 2,876 cases and 81 deaths.
- Gabon confirmed its first case.
- Germany's case total rose to 3,675.

- Ghana confirmed its first two cases, a Norwegian Embassy official and a Turkish citizen.
- Guadeloupe confirmed its first case, a citizen who recently returned from France.
- Guatemala confirmed its first case, a traveller from Italy.
- Guinea reported its first case, an employee of the EU delegation.
- Italy's cases rose to 17,660 and deaths rose to 1,266, a rise of 250 deaths in 24 hours.
- Kazakhstan confirmed its first two cases, which were also the first in Central Asia.
- Kenya confirmed its first case, a Kenyan national who had returned from the United States via London.
- Kosovo confirmed its first two cases.
- Lithuania confirmed three new cases, bringing the total number to 6. A Spanish citizen from Madrid is confirmed in Vilnius, along with a woman in Klaipėda who arrived from Tenerife, and a man in Kaunas who arrived from Italy on 29 February.
- Malaysia confirmed 39 new cases, bringing the total number to 197.
- The Philippines confirmed 12 more cases, bringing the total number to 64.
- Panama confirmed several new cases, bringing the total to 27.
- Puerto Rico confirmed its first three cases, a 71-year-old man and an Italian couple, aged 68 and 70.
- Saint Lucia confirmed its first case, a 63-year-old woman who had traveled to the UK.
- Singapore confirmed 13 more cases, bringing the total number to 200. The number of imported cases surpassed that of local cases.
- Slovakia confirmed 11 new cases, bringing the total number to 32.
- Spain's cases rose to 4,231. 35 more deaths were also confirmed, bringing the total to 121.
- Sri Lanka confirmed three new cases, bringing the total number to 6.
- Sudan confirmed its first case and first death, a man in his 50s who travelled to the United Arab Emirates.
- Suriname confirmed its first case.
- Switzerland reported 1,125 cases (of which 116 are undergoing final analysis), a 31% day-on-day increase.
- Panama confirmed several new cases, bringing the total number to 36.

- Turkey confirmed its second case. Later in the day, the case count increased to 5. All cases are related to the first case, who contracted the virus from Europe.
- Ukraine reported its first death, a 71-year-old woman from Radomyshl, Zhytomyr Oblast who recently travelled to Poland.
- The United Kingdom confirmed 202 new cases, bringing the total number to 798. The first death was confirmed in Scotland, bringing the total deaths across the UK to 11.
- The United States confirmed 559 additional cases, bringing the total number to 2,204. Eight more deaths were reported, bringing the total number to 49.
- U.S. Virgin Islands confirmed the first case in the territory.
- Uruguay confirmed its first four cases, all of them having travelled from Milan, Italy.
- Venezuela confirmed its first two cases, one a traveller from the United States, and the second who had travelled from Spain.
- Vietnam confirmed three additional cases, bringing the total number to 47.

14 March

- Brazil confirmed seven more cases, bringing the total number to 178.
- Brunei reported three new cases, bringing the total to 40.
- Bulgaria confirmed six new cases bringing the total number to 37. One more death was also confirmed, bringing the total number to 2.
- Central African Republic confirmed its first case.
- Colombia reported additional eight more cases, taking the total to 24.
- Congo Republic confirmed its first case, a person who had travelled from France.
- Denmark confirmed the first death in the country, an 81-year-old man.
- Equatorial Guinea confirmed its first case, a 42-year-old woman who returned from Madrid.
- Eswatini confirmed its first case, a 33-year-old woman, who traveled to the United States and then Lesotho before returning home to Eswatini.
- France confirmed 838 new cases, bringing the total number to 4,499. 12 more deaths were confirmed, bringing the total to 91.

- Iran confirmed 1,365 new cases, bringing the total number to 12,729. 97 more deaths were confirmed, bringing the total to 611.
- Ireland confirmed 39 new cases, the largest to date, and one more confirmed death. 129 total confirmed cases and 2 deaths.
- Italy confirmed 3,497 new cases, bringing the total number to 21,157. 175 more deaths were also confirmed, bringing the total to 1,441.
- Malaysia confirmed 41 new cases, bringing the total number to 238.
- Mauritania confirmed its first case.
- Mayotte confirmed its first case.
- Moldova confirmed 4 new cases.
- Namibia confirmed its first cases, two tourists visiting the country.
- New Zealand confirmed its sixth case, a man who had returned from the United States on 6 March.
- The Philippines confirmed 47 more cases, bringing the total number to 111.
- Rwanda confirmed its first case.
- Saudi Arabia confirmed 17 additional cases, with its total rising to 103.
- Singapore confirmed 12 more cases, bringing the total number to 212.
- Seychelles reported its first two cases.
- Spain confirmed 1,522 new cases, raising its total to 5,753 cases. 62 more deaths were confirmed, bringing the total to 183.
- Sri Lanka confirmed two new cases, bringing the total number to 8.
- Another three cases were later confirmed, bringing the total number to 11.
- Panama confirmed several new cases, bringing the total number to 43.
- Thailand confirmed seven more cases, bringing the total number to 82.
- The United Kingdom confirmed 342 new cases, bringing the total number to 1,140. 10 more deaths were reported, bringing the total to 21.
- The United States confirmed 612 new cases, bringing the total number to 2,816. One of the cases was 24-year-old Christian Wood of the Detroit Pistons. Eleven more deaths were reported, bringing the total number to 60.
- Vietnam confirmed six additional cases, bringing the total number to 53.

15 March

- Akrotiri and Dhekelia bases in Cyprus announced the first two cases among British military personnel.
- Bahamas confirmed their first case, a person with no recent travel history.
- Brunei confirmed ten new cases, bringing the total number to 50.
- Bulgaria confirmed eight new cases, bringing the total number to 51.
- Finland ceased testing for people returning from trips abroad, and all of the people suffering flu symptoms in the country. The tests are reserved for health professionals only.
- Guam confirmed its first three cases, two people who arrived from Manila and one other person with no recent travel history.
- Indonesia confirmed 21 more cases, bringing the total number to 117. Among the new diagnoses include North Sulawesi and Yogyakarta's first cases.
- Ireland confirmed 40 new cases, the largest to date. 169 total confirmed cases and 2 deaths.
- Italy confirmed 3,590 more cases, bringing the total number to 24,747. The country also recorded 368 new deaths, bringing the total number to 1,809. This was the largest death rate in a day for a country since the pandemic started.
- Malaysia confirmed 190 new cases, bringing the total number to 428. Most of these cases are linked to a gathering in Kuala Lumpur.
- Moldova confirmed 11 new cases, bringing the total number to 23.
- New Zealand confirmed two new cases, a Wellington man and a Danish woman, bringing the total to 8.
- Saudi Arabia confirmed 15 new cases, bring the total number to 118
- Singapore confirmed 14 new cases, bringing the total number to 226. This is the highest number of new cases in the country in a single day.
- Sri Lanka confirmed eight more cases, bringing the total to 19. There are 18 active cases with one recovered case.
- Panama confirmed several new cases, bringing the total number to 55.
- Thailand confirmed 32 more cases, bringing the total number to 114.
- Turkey confirmed 12 more cases, bringing the total number to 18.
- The United Kingdom confirmed 232 new cases, bringing the total number to 1,372. 14 more deaths were reported, bringing the total to 35.

- The United States confirmed 669 new cases, bringing the total number to 3,485. Five more deaths were reported, bringing the total number to 65.
- Uzbekistan confirmed its first case, a citizen who had returned from France.
- Vietnam confirmed four additional cases, bringing the total number to 57.

16 March

- Benin confirmed its first case, a 49-year-old man who had traveled to Belgium and Burkina Faso.
- Bulgaria confirmed two new cases, bringing the total number to 53. Bulgaria later confirmed nine more cases, bringing the final total to 62.
- Costa Rica confirmed 40 cases.
- Cyprus confirmed 13 new cases, taking the total to 46.
- Egypt confirmed 40 new cases, taking the total to 166. It also confirmed two additional deaths.
- France confirmed that cases had risen to 6,633 with 148 deaths.
- Germany's cases rose to 7,272 with 17 deaths.
- Greece confirmed 21 new cases for a total of 352.
- Greenland confirmed its first case.
- Guatemala confirmed five more cases, bringing the total to 6 and the first death in the country from the virus.
- Indonesia confirmed 17 more cases, 134 in total.
- Iran's cases rose to 13,938 with its death toll rising by 129 to 853. The deaths included Hashem Bathaie Golpayenagi, Grand Ayatollah and a representative of the Tehran Province in Iran's Assembly of Experts.
- Ireland confirmed 54 new cases, the largest to date. 223 total confirmed cases and 2 deaths.
- Italy's confirmed cases rose to 27,980 and deaths to 2,158. This total included 3,233 new cases and 349 new deaths.
- Jordan confirmed four new cases, taking its total to 15.
- Kuwait confirmed 11 new cases, taking the total to 123.
- Liberia confirmed its first case.

- Malaysia announced another 125 cases, taking it to a total of 553 with many of the new cases linked to a recent religious festival in the country.
- Moldova announced another six cases, taking its total to 29.
- Oman confirmed two additional cases, raising the total to 24.
- Qatar reported 64 new cases taking it to 401.
- Singapore confirmed 17 more cases, bringing the total number to 243. This is the biggest spike of new cases in a single day.
- Somalia confirmed its first case.
- Spain reported 9,942 cases and 342 deaths.
- Sri Lanka confirmed eleven more cases, bringing the total to 29.
- Tanzania confirmed its first case.
- Turkey announced that the number of cases had risen to 47, with 29 new cases confirmed.
- The United Kingdom confirmed a total of 1,543 positive tests, up from 1,372.
- The United States confirmed 974 new cases, bringing the total number to 4,459. 22 more deaths were reported, bringing the total number to 87.
- Ukraine confirmed two new cases: two women, one of whom recently returned from Italy. Later in the day, two more cases were confirmed, bringing the total to 7.
- Vietnam confirmed four additional cases, bringing the total number to 61.

17 March

- Barbados reported its first two cases, people who recently returned from the US.
- Bulgaria confirmed five new cases, bringing the number to 67, and then an additional 14 new cases, bringing the total number to 81.
- Brazil confirmed the first death in the country, a 62-year-old man in the state of São Paulo.
- France confirmed 7,730 people had now tested positive with 175 deaths.
- Gambia reported its first case.
- Hungary reported 11 new cases, bringing the total number to 50.
- Indonesia reported 38 new cases, bringing the total number to 172.

- Ireland confirmed 69 new cases, the largest to date. 292 total confirmed cases and 2 deaths.
- Italy confirmed an increase to 31,506 cases and 2,503 deaths.
- Kuwait confirmed seven new cases, taking the total to 130.
- Malaysia reported its first two deaths, a 60 year old pastor in Sarawak and a 34-year-old man from Johor who had attended a Muslim religious gathering. 120 new cases were also confirmed, bringing the total number to 673.
- Montenegro reported its first two cases. One infected was from Podgorica, and the other from Ulcinj. Montenegro was the last country in Europe without any confirmed cases.
- The Netherlands confirmed 1,705 cases and 43 deaths.
- New Zealand confirmed four new cases, bringing the total to 12.
- Pakistan confirmed its first death.
- The Philippines reported 45 new cases, taking the total to 187.
- Singapore confirmed 23 new cases, bringing the total number to 266. This is the highest number of new cases in the country in a single day so far.
- Sint Maarten confirmed its first case, a 26-year-old local resident who traveled to UK and Miami.
- Slovakia confirmed 25 new cases, bringing the total number to 97.
- Panama confirmed several new cases, bringing the total number to 86.
- Ukraine reported seven new cases and its second death, bringing the total number of cases to 14.
- The United Kingdom reported 407 new cases, with the total rising from 1,543 to 1,950.
- The United States confirmed 1,676 new cases, bringing the total number to 6,135. 25 more deaths were reported, bringing the total number to 112. The Brooklyn Nets announced that four of their players had tested positive, including forward Kevin Durant.
- Vietnam confirmed five more cases, bringing the total number to 66.

18 March

- Bermuda confirmed its first two cases.
- Bangladesh confirmed its first death.
- Costa Rica confirmed its first death.
- Djibouti confirmed its first case.

- El Salvador confirmed its first case.
- Guatemala confirmed two additional cases, bringing the total to 8.
- Ireland confirmed 74 new cases, the largest to date. 366 total confirmed cases and 2 deaths.
- Italy reported 4,207 new cases and 475 new deaths. The total number of cases reached 35,713 and the total death toll rose to 2,978.
- Kyrgyzstan reported its first cases, three people who had recently returned from a pilgrimage to Saudi Arabia.
- Malaysia confirmed 117 new cases, bringing the total number to 790.
- Mauritius announced its first three cases.
- Moldova confirmed its first death and six new cases, bringing the total to 36.
- Montserrat confirmed its first case, a person who visited the United Kingdom.
- New Caledonia confirmed its first two cases.
- New Zealand confirmed eight new cases, bringing the total to 20.
- Nicaragua confirmed its first case, imported from Panama.
- Singapore confirmed 47 new cases, bringing the total number to 313. This is the highest number of new cases in the country in a single day so far.
- Slovakia confirmed its first death.
- Panama confirmed several new cases, bringing the total to 109.
- The United Kingdom confirmed 676 new cases, taking it to a total of 2,626. 32 additional deaths were announced, taking the total to 104.
- The United States confirmed 2,601 more cases, bringing the total number to 8,736. 37 more deaths were reported, bringing the total number to 149.
- Vietnam confirmed ten more cases, bringing the total number to 76.
- Zambia reported its first two cases.

19 March

- Angola confirmed its first case, a Chinese businessman.
- Chad confirmed its first case, a Moroccan national who had traveled to the country from Cameroon.
- Fiji confirmed its first case.
- Guatemala confirmed one additional case, raising the total to 9.
- Haiti reports its first two cases.

- Ireland confirmed 191 new cases, the largest to date with more than two and a half times the confirmed cases of 18 March, and a third death, bringing the total number of cases to 557. Approximately 7,000 tests have been administered to date and the Minister for Health, Simon Harris, believes there will be 15,000 tests given a day starting within the next few days.
- Isle of Man confirmed its first case, a person who recently returned from Spain.
- Italy's number of cases rose to 41,035 from a previous 35,713, up by 5,322, a faster rate of growth than the past three days. They also overtook China as the country with the most deaths, registering 3,405 dead, a rise of 427 from the day before.
- Malaysia confirmed 110 new cases, bringing the total number to 900.
- New Zealand confirmed eight new cases, bringing the total to 28.
- Niger confirmed its first case, a local 36-year-old man who traveled a lot to Togo, Ghana, Ivory Coast and Burkina Faso due to work.
- Panama reported several new cases, bringing the total number to 137.
- Pakistan confirmed 70 additional cases.
- Russia confirmed its first death.
- Singapore confirmed 32 new cases, bringing the total number to 345.
- Spain reported a large increase in cases, with 17,147 cases and 767 deaths.
- The United Kingdom's total of diagnosed cases rose to 3,229, an increase of 603, with the death toll rising to 144.
- The United States confirmed 4,397 more cases, bringing the total number to 13,133. 46 more deaths were reported, bringing the total number to 195.
- The Denver Nuggets, Los Angeles Lakers, Philadelphia 76ers, and the Boston Celtics all reported that some of their players had tested positive for the virus. The players had not been named.
- Vietnam confirmed nine more cases, bringing the total number to 85.

20 March

- Cape Verde confirmed its first case, a 62-year-old English tourist.
- East Timor confirmed its first case.
- Guatemala confirmed three additional cases, bringing the total to 12.

- Ireland confirmed 126 new cases, less than that of 19 March. 683 total confirmed cases and 3 deaths.
- Madagascar reported its first three cases.
- Malaysia confirmed its third death. In addition, 130 new cases were confirmed, bringing the total number to 1,030.
- New Zealand confirmed 11 new cases, bringing the total to 39.
- Panama reported over 60 new cases, bringing the total number to 200.
- Papua New Guinea confirmed its first case.
- Singapore confirmed 40 new cases, bringing the total number to 385.
- Uganda confirmed its first case.
- Ukraine confirmed 15 new cases, bringing the total number to 41.
- The United Kingdom confirmed that positive test results had risen to 3,983 and there had been 177 deaths.
- The United States confirmed 5,630 more cases, bringing the total number to 18,763. 63 more deaths were reported, bringing the total number to 258.
- Vietnam confirmed six more cases, bringing the total number to 91.
- Zimbabwe confirmed its first case.

21 March

- Åland Islands confirmed their first two cases.
- Eritrea confirmed its first case, a 39-year-old Eritrean national with permanent residence in Norway.
- Guatemala confirmed five new cases, bringing the total to 17.
- Indonesia confirmed 81 new cases, bringing the total to 450. The number of deaths rose by six to 38.
- Ireland confirmed 102 new cases, less than that of 20 March, with a total of 785 total confirmed cases and 3 deaths.
- Malaysia confirmed five more deaths, bringing the total to 8. Separately, 153 new cases were confirmed, bringing the total number to 1,183.
- New Zealand reported 13 new cases, bringing the total to 52.
- The Philippines recorded 77 new cases, bringing the total number to 307.
- Singapore confirmed its first two deaths, a 64-year-old Indonesian man and a 75-year-old Singaporean woman. 47 new cases were later confirmed, bringing the total number to 432.

- Transnistria confirmed its first two cases.
- The United Kingdom confirmed that positive test results had risen to 5,018 and there had been 233 deaths.
- The United States confirmed 4,886 more cases, bringing the total number to 23,649. 44 more deaths were reported, bringing the total number to 302.
- Vietnam confirmed three more cases, bringing the total number to 94.

22 March

- Dominica confirmed its first case, a 54-year-old man who recently arrived from the United Kingdom.
- Grenada confirmed its first case, a woman who recently travelled to the UK.
- Indonesia confirmed 64 new cases, bringing the total number to 514.
- Ireland confirmed 121 new cases, more than that of 21 March, and a fourth death, with a total of 906 total confirmed cases and 4 deaths.
- Malaysia confirmed two more deaths, bringing the total to 10. There were also 123 new cases confirmed, bringing the total number to 1,306.
- Mozambique confirmed its first case.
- New Zealand reported 14 new cases, bringing the total to 66.
- Romania confirmed its first death, a 67-year-old man suffering from cancer. A total of 367 cases have since been recorded.
- Singapore confirmed 23 new cases, bringing the total number to 455.
- Syria recorded its first case.
- The United Kingdom confirmed that positive test results had risen to 5,683 and there had been 281 deaths.
- The United States confirmed 10,351 more cases, bringing the total number to 34,000. 111 more deaths were reported, bringing the total number to 413.
- Ukraine confirmed 26 more cases, bringing the total number to 73.
- Vietnam confirmed 19 more cases, bringing the total number to 113.

23 March

- Belize confirmed its first case, a local resident who recently returned from Los Angeles.
- Indonesia confirmed 65 new cases, bringing the total number to 579.

- Ireland confirmed 219 new cases, the largest increase to date, and two deaths, with a total of 1,125 total confirmed cases and 6 deaths.
- Myanmar confirmed its first two cases.
- New Zealand confirms 36 new cases, bringing the total number to 102.
- Panama confirmed several new cases, bringing the total number to 345. Panama also confirmed several new deaths, bringing the total number to 6.
- Singapore confirmed 54 new cases, bringing the total number to 509.
- Turks and Caicos Islands confirm their first case.
- The United Kingdom confirmed that positive test results had risen to 6,650 and there had been 335 deaths.
- Vietnam confirmed 10 more cases, bringing the total number to 123.

24 March

- Easter Island reports its first case.
- Indonesia confirmed 107 new cases, the largest increase it has had to date, bringing the total number of cases to 686.
- Italy confirmed 5,249 new cases and 743 new deaths, taking the country to a total of 69,176 cases and 6,820 deaths.
- Laos reports its first two cases.
- Libya confirms its first case.
- Malaysia reported 106 new cases, bringing the total number to 1,624. Malaysia also reported one death, bringing the total fatality rate to 15.
- New Caledonia reports two new cases, bringing the total to nine.
- New Zealand confirmed 40 new cases, and started including probable cases in its count. The inclusion brings New Zealand's total to 155.
- Panama confirmed 98 new cases, bringing the total number to 443, and 2 deaths, bringing the total number to 8.
- Singapore confirmed 49 new cases, bringing the total number to 558.
- The United Kingdom confirmed that positive test results had risen to 8,077 and there had been 422 deaths. The 8,077 cases included 6,843 in England, 584 in Scotland, 478 in Wales and 172 in Northern Ireland.

****

Chapter – 7

Responses about COVID-19

Domestic responses

China

"Aerial photography of roads after motor vehicles are banned in central urban areas of Wuhan: few vehicle traces"—Video news from China News Service.

The first person known to have fallen ill due to the new virus was traced back to 1 December 2019 in Wuhan. Doctor Zhang Jixian observed a cluster of unknown pneumonia on 26 December, and her hospital informed Wuhan Jianghan CDC on 27 December. A public notice on the outbreak was released by Wuhan Municipal Health Commission on 31 December. WHO was informed of the outbreak on the same day. At the same time these notifications were happening, doctors in Wuhan were being threatened by policy for sharing information about the outbreak. Chinese National Health Commission initially said that they had no "clear evidence" of human-to-human transmissions.

Temporary hospital for treating mild COVID-19 patients in Wuhan, converted from Tazihu Sports Centre. There were more than 10 temporary hospitals like this in Wuhan. Each of them could accommodate thousands of mild patients. After these hospitals had come into operation, the epidemic in Wuhan eased.

The Chinese Communist Party launched a radical campaign described by the Party general secretary Xi Jinping as a "people's war" to contain the spread of the virus. In what has been described as "the largest quarantine in human history", a quarantine was announced on 23 January stopping travel in and out of Wuhan, which was extended to a total of 15 cities in Hubei, affecting a total of about 57 million people. Private vehicle use was banned in the city. Chinese New Year (25 January) celebrations were cancelled in many places. The authorities also announced the construction of a temporary hospital, Huoshenshan Hospital, which was completed in 10 days, and 14 temporary hospitals were constructed in China in total.

On 26 January, the Communist Party and the government instituted further measures to contain the COVID-19 outbreak, including health declarations for travellers and changes to national holidays. The leading group decided to extend the Spring Festival holiday to contain the outbreak. Universities and schools around the country were also closed. The regions of Hong Kong and Macau instituted several measures, particularly in regard to schools and universities. Remote working measures were instituted in several Chinese regions. Travel restrictions were enacted. Other provinces and cities outside Hubei imposed travel restrictions. Public transport was modified, and museums throughout China were temporarily closed. Control of movement of people was applied in many cities, and it has been estimated that about 760 million people (more than half the population) faced some form of outdoor restriction.

After the outbreak entered its global phase in March, many Chinese students studying in Europe and the United States have returned home as the domestic daily new cases in China declined. Chinese authorities have taken strict measures to prevent the virus from "importing" from other countries. For example, Beijing has imposed a 14-day mandatory quarantine for all international travellers entering the city.

The early response by the Wuhan authorities was criticized as prioritizing control of information that might be unfavourable for local officials over public safety, and the Chinese government was also criticized for cover-ups and downplaying the initial discovery and severity of the outbreak. In early January 2020, Wuhan police summoned and "admonished" several doctors—including Li Wenliang, an ophthalmologist at Wuhan Central Hospital—for "spreading rumours" likening the disease to SARS. Li later died because of the virus. Later in March, Wuhan police apologized to Li's family after National Supervisory Commission admitted the conduct of local officials was inadequate and acknowledged the whistleblower's effort on raising public awareness. Observers have also blamed the institutional censorship that left the citizens and senior officials with inaccurate information on the outbreak and "contributed to a prolonged period of inaction that allowed the virus to spread". Some experts doubted the accuracy of the number of cases reported by the Chinese government, which repeatedly changed how it counted coronavirus cases, while others say it wasn't likely a deliberate attempt to manipulate the data. The Chinese government has also been accused of rejecting help from the US CDC and the WHO.

Although criticisms have been levelled at the aggressive response of China to control the outbreak, China's actions have also been praised by some foreign leaders such as US President Donald Trump, and Russian president Vladimir Putin. Trump later reversed himself, saying "I wish they could have told us earlier about what was going on inside," adding that China "was very secretive, and that's unfortunate". The director of WHO Tedros Adhanom Ghebreyesus congratulated Chinese government "for the extraordinary measures it has taken to contain the outbreak", and a later WHO report described China's response as "perhaps the most ambitious, agile and aggressive disease containment effort in history". According to a media report on 16 March, the economy in China was very hard hit in the first two months of 2020 due to the measures taken by the government to curtail virus spread, and retail sales plunged 20.5%. Per media reports, on 23 March mainland China has gone five days with only one case transmitted domestically, in this instance via a traveller returning to Guangzhou from Istanbul.

On 24 March 2020, Chinese Premier Li Keqiang reported that spread of domestically transmitted epidemic has been basically blocked and the outbreak has been controlled in China.

South Korea

COVID-19 was confirmed to have spread to South Korea on 20 January 2020 from China. There was a large increase in cases on 20 February, potentially attributable to a gathering in Daegu of a new religious movement known as the Shincheonji Church of Jesus. Shincheonji devotees visiting Daegu from Wuhan were suspected to be the origin of the outbreak. As of 22 February, among 9,336 followers of the church, 1,261 or about 13% reported symptoms.

South Korea declared the highest level of alert on 23 February 2020. On 28 February, more than 2,000 confirmed cases were reported in Korea, rising to 3,150 on 29 February. All South Korean military bases were on quarantine after tests confirmed that three soldiers were positive for the virus. Airline schedules were also affected and therefore they were changed.

South Korea introduced what was considered the largest and best-organised program in the world to screen the population for the virus, and isolate any infected people as well as tracing and quarantining those who contacted them. Screening methods included a drive-thru testing for the virus with the results available the next day. It is considered to be a success in controlling the outbreak despite not quarantining entire cities.

The South Korean society was initially polarised with President Moon Jae-in's response to the crisis. Many Koreans signed petitions either calling for the impeachment of Moon over what they claimed is the government's mishandling of the outbreak, or praising his response. On 23 March, it was reported that South Korea had the lowest one-day case total in four weeks. South Korea's approach to the outbreak includes having 20,000 people tested every day for coronavirus.

Iran

Iran reported its first confirmed cases of SARS-CoV-2 infections on 19 February in Qom, where, according to the Ministry of Health and Medical Education, two people had died later that day. Early measures announced by the government included the cancellation of concerts and other cultural events, sporting events, and Friday prayers, universities, higher education institutions and schools. Iran allocated five trillion rials to combat the virus. President Hassan Rouhani said on 26 February 2020 that there were no plans to quarantine areas affected by the outbreak, and only individuals would be quarantined. Plans to limit travel between cities were announced in March, although heavy traffic between cities ahead of the Persian New Year Nowruz continued. Shia shrines in Qom remained open to pilgrims until 16 March 2020.

Iran became a centre of the spread of the virus after China. Amidst claims of a cover-up of the extent of the outbreak in the country, more than ten countries had traced their cases back to Iran by 28 February, indicating that the extent of the outbreak may be more severe than the 388 cases reported by the Iranian government by that date. The Iranian Parliament was shut down, with 23 of its 290 members reported to have had tested positive for the virus on 3 March. At least 12 sitting or former Iranian politicians and government officials had died from the disease by 17 March 2020. On 15 March, the Iranian government reported 100 deaths in a single day, the most recorded since the outbreak began. Per media reports on 23 March Iran has 50 new cases every hour and one new death every ten minutes due to coronavirus. Even so, some sources like Radio Farda, which is US backed, says Iran may be underreporting.

Europe

As of 18 March, more than 250 million people are in lockdown in Europe.

Italy

Civil Protection volunteers carrying out health checks at the Guglielmo Marconi Airport in Bologna.

The outbreak was confirmed to have spread to Italy on 31 January, when two Chinese tourists tested positive for SARS-CoV-2 in Rome. Cases began to rise sharply, which prompted the Italian government to suspend all flights to and from China and declare a state of emergency. An unassociated cluster of COVID-19 cases was later further detected starting with 16 confirmed cases in Lombardy on 21 February.

On 22 February, the Council of Ministers announced a new decree-law to contain the outbreak, including quarantining more than 50,000 people from 11 different municipalities in northern Italy. Prime Minister Giuseppe Conte said, "In the outbreak areas, entry and exit will not be provided. Suspension of work activities and sports events has already been ordered in those areas."

On 4 March, the Italian government ordered the full closure of all schools and universities nationwide as Italy reached 100 deaths. All major sporting events, including Serie A football matches, will be held behind closed doors until April. On 9 March, all sport was suspended completely for at least one month. On 11 March, Prime Minister Conte ordered stoppage of nearly all commercial activity except supermarkets and pharmacies.

On 6 March, the Italian College of Anaesthesia, Analgesia, Resuscitation and Intensive Care (SIAARTI) published medical ethics recommendations regarding triage protocols that might be employed. On 19 March, Italy overtook China as the country with the most coronavirus-related deaths in the world after reporting 3,405 fatalities from the pandemic. As of 24 March 2020, there were 69,176 confirmed cases, 6,820 deaths and 8,326 recoveries in Italy. On the same date it was reported that Russia had sent nine military planes with medical equipment to Italy.

Spain

People maintain social distancing in Valencia, Spain

The 2019–20 coronavirus pandemic was confirmed to have spread to Spain on 31 January 2020, when a German tourist tested positive for SARS-CoV-2 in La Gomera, Canary Islands.

On 24 February, following a COVID-19 outbreak in Italy, Spain confirmed multiple cases related to the Italian clusters, originating from a medical doctor from Lombardy, Italy, who was on holiday in Tenerife. Afterwards, multiple COVID-19 cases were detected in Tenerife involving people who had come in contact with the doctor. Other cases involving individuals who visited Italy were also discovered in mainland Spain.

On 24 March, it was reported that there had been 41,969 cases, and that 514 people had died in a single day. As of 21 March 2020, at least 2,355 people were hospitalised in intensive care (942 in the Community of Madrid) and more than 350,000 tests for COVID-19 had been conducted. Since 13 March 2020, there have been registered cases in all 50 provinces of the country. Only on the islands of La Graciosa (Canary Islands) and Formentera (Balearic Islands) have there been no registered positive cases.

United Kingdom

People in London buying canned foods and toilet paper on 18 March 2020

The UK response to the virus first emerged as one of the most relaxed of the affected countries, and until 18 March 2020, the British government did not impose any form of social distancing or mass quarantine measures on its citizens. As a result, the government received criticism for the perceived lack of pace and intensity in its response to concerns faced by the public.

On 16 March, Prime Minister Boris Johnson made an announcement advising against all non-essential travel and social contact, to include working from home where possible and avoiding venues such as pubs, restaurants and theatres. On 20 March, the government announced that all leisure establishments (pubs, gyms etc.) were to close as soon as possible, and promised to pay up to 80% of workers' wages, to a limit of £2,500 per month, to prevent unemployment in the crisis.

On 23 March, the Prime Minister announced tougher social distancing measures, banning gatherings of more than two people and restricting travel and outdoor activity to that deemed strictly necessary. Unlike previous measures, these restrictions were enforceable by police through the issuing of fines and the dispersal of gatherings. Most businesses were ordered to close, with "essential" exceptions including supermarkets, pharmacies, banks, hardware shops, petrol stations and garages.

France

The virus was confirmed to have spread to France on 24 January 2020, when the first COVID-19 case in Europe and France was confirmed in Bordeaux. It involved a 48-year-old French citizen who arrived in France from China. Two more cases were confirmed by the end of the day; all of the individuals recently returned from China. A Chinese tourist was admitted to a hospital in Paris on 28 January and died on 14 February, marking the first death from COVID-19 in Europe and France. It was also the first death outside of Asia. A national lockdown was put in place on 27 March. As of 23 March 2020, there have been 19,856 confirmed cases, 860 deaths and at least 2,200 recoveries in France.

Germany

The virus was confirmed to have been transmitted to Germany on 27 January 2020, when the first COVID-19 case was confirmed in Bavaria. The majority of the COVID-19 cases in January and early February originated from the headquarters of a car parts manufacturer in Bavaria. Later, new clusters were introduced by travellers from Italy, China and Iran. As of 24 March 2020, Germany has reported 32,035 cases, 150 deaths and 3,133 recoveries.

Rest of Europe

Despite a select number of countries being particularly affected by bad outbreaks of COVID-19, many European countries, as of 24 March experienced far fewer cases of the disease. These include Poland, Lithuania, Finland, Greece, Croatia and Armenia among many others.

United States

The first known case in the United States of COVID-19 was confirmed in the Pacific Northwest state of Washington on 20 January 2020, in a man who had returned from Wuhan on 15 January. The White House Coronavirus Task Force was established on 29 January. On 31 January, the Trump administration declared a public health emergency, and placed travel restrictions on entry for travellers from China.

President Trump signs the Coronavirus Preparedness and Response Supplemental Appropriations Act into law on 6 March 2020

After the first death in the United States was reported in Washington state on 29 February, its governor, Jay Inslee, declared a state of emergency, an action that was followed by other states. Schools in the Seattle area cancelled classes on 3 March, and by mid-March, schools across the country were closing and most of the country's students were out of school.

On 6 March, President Donald Trump signed the Coronavirus Preparedness and Response Supplemental Appropriations Act, which provided $8.3 billion in emergency funding for federal agencies to respond to the outbreak. Corporations imposed employee travel restrictions, cancelled conferences, and encouraged employees to work from home. Sports events and seasons were cancelled.

On 11 March, Trump announced travel restrictions for most of Europe (excluding the United Kingdom) for 30 days, effective 13 March, and on 14 March, he expanded the restrictions to include the United Kingdom and Ireland. On 13 March, he declared a national emergency, which made federal funds available to respond to the crisis. Beginning on 15 March, many businesses closed or reduced hours throughout the US as a method to try to combat the virus.

As of 24 March 2020, the epidemic was present in all 50 states, plus the District of Columbia. The number of confirmed cases in the US rose to 54,823, with 778 deaths. On 23 March, it was reported that New York city had 10,700 cases of the coronavirus, an amount that is greater than the country of South Korea.

The White House has been criticized for downplaying the threat and controlling the messaging by directing health officials and scientists to coordinate public statements and publications related to the virus with the office of Vice President Mike Pence. Overall approval of Trump's management of the crisis has been polarized along partisan lines.

International responses

An analysis of air travel patterns was used to map out and predict patterns of spread and was published in The Journal of Travel Medicine in mid-January 2020. Based on information from the International Air Transport Association (2018), Bangkok, Hong Kong, Tokyo and Taipei had the largest volume of travellers from Wuhan. Dubai, Sydney and Melbourne were also reported as popular destinations for people travelling from Wuhan. Bali was reported as least able in terms of preparedness, while cities in Australia were considered most able.

Australia released its Emergency Response Plan for Novel Coronavirus (COVID-19) on 7 February. It states that much is yet to be discovered about COVID-19, and that Australia will emphasize border control and communication in its response to the pandemic.

A worker disinfects an ambulance in Daegu, South Korea

Announcement on a television screen at Delft University of Technology in the Netherlands, advising students not to travel to China (19 February 2020)

Travel restrictions

Countries that have implemented a global travel ban in response to the COVID-19 pandemic.

As a result of the outbreak, many countries and regions including most of the Schengen Area, Armenia, Australia, India, Iraq, Indonesia, Kazakhstan, Kuwait, Malaysia, Maldives, Mongolia, New Zealand, Philippines, Russia, Singapore, Sri Lanka, Taiwan, Vietnam, and the United States have imposed temporary entry bans on Chinese citizens or recent visitors to China, or have ceased issuing visas or reimposed visa requirements on Chinese citizens. Samoa started refusing entry to its own citizens who had previously been to China, attracting widespread condemnation over the legality of the decision.

The European Union rejected the idea of suspending the Schengen free travel zone and introducing border controls with Italy, which has been criticized by some European politicians. After some EU member states announced complete closure of their national borders to foreign nationals, the European Commission President Ursula von der Leyen said that "Certain controls may be justified, but general travel bans are not seen as being the most effective by the World Health Organization."

Saudi Arabia temporarily banned foreigners from entering Mecca and Medina, two of Islam's holiest pilgrimage sites, to prevent the spread of coronavirus. The United States suspended travel from the Schengen Area and later the Common Travel Area. Many countries then started closing their borders to virtually all non-citizens or non-residents in rapid succession, including India, Slovakia, Denmark, Poland, Lithuania, Oman, United Arab Emirates and Russia.

Evacuation of foreign citizens

Ukraine evacuates Ukrainian and foreign citizens from Wuhan

Owing to the effective quarantine of public transport in Wuhan and Hubei, several countries have planned to evacuate their citizens and diplomatic staff from the area, primarily through chartered flights of the home nation, with Chinese authorities providing clearance. Canada, the United States, Japan, India, France, Australia, Sri Lanka, Germany and Thailand were among the first to plan the evacuation of their citizens. Pakistan has said that it will not be evacuating any citizens from China. On 7 February, Brazil evacuated 34 Brazilians or family members in addition to four Poles, a Chinese person and an Indian citizen. The citizens of Poland, China and India deplaned in Poland, where the Brazilian plane made a stopover before following its route to Brazil. Brazilian citizens who went to Wuhan were quarantined at a military base near Brasília. On the same day, 215 Canadians (176 from the first plane, and 39 from a second plane chartered by the US government) were evacuated from Wuhan to CFB Trenton to be quarantined for two weeks.

On 11 February, another plane of 185 Canadians from Wuhan landed at CFB Trenton. Australian authorities evacuated 277 citizens on 3 and 4 February to the Christmas Island Detention Centre, which had been repurposed as a quarantine facility, where they remained for 14 days. A New Zealand evacuation flight arrived in Auckland on 5 February; its passengers (including some from Australia and the Pacific) were quarantined at a naval base in Whangaparoa, north of Auckland. The United States announced that it would evacuate Americans aboard the cruise ship Diamond Princess. On 21 February, a plane carrying 129 Canadian passengers who had been evacuated from Diamond Princess landed in Trenton, Ontario. The Indian government has scheduled its air force to evacuate its citizens from Iran.

International aid

Digital billboard conveying support with the words "Be Strong China" in various languages at Shibuya in Tokyo on 10 February

On 5 February, the Chinese foreign ministry stated that 21 countries (including Belarus, Pakistan, Trinidad and Tobago, Egypt and Iran) had sent aid to China. The US city of Pittsburgh announced plans to send medical aid to Wuhan, which is its sister city. The University of Pittsburgh Medical Center (UPMC) also announced plans to provide help. Some Chinese students at American universities have joined together to help send aid to virus-stricken parts of China, with a joint group in the greater Chicago area reportedly managing to send 50,000 N95 masks and 1,500 protection suits to hospitals in the Hubei province on 30 January.

The humanitarian aid organisation Direct Relief, in coordination with FedEx transportation and logistics support, sent 200,000 face masks along with other personal protective equipment, including gloves and gowns, by emergency airlift to the Wuhan Union Hospital by 30 January. The Gates Foundation stated on 26 January that it would donate US$5 million in emergency funds to support the response in China, along with technical support for front-line responders. On 5 February, Bill and Melinda Gates further announced a US$100 million donation to the WHO to fund vaccine research and treatment efforts along with protecting "at-risk populations in Africa and South Asia".

Japan, in planning a flight to Wuhan to pick up Japanese nationals there, promised that the plane would bring aid supplies that, according to Japanese foreign minister Toshimitsu Motegi, would consist of "masks and protective suits for Chinese people as well as for Japanese nationals". On 26 January, the plane arrived in Wuhan, donating its supply of one million face masks to the city. Among the aid supplies were 20,000 protective suits for medical staff across Hubei donated by the Tokyo Metropolitan Government.

Tehran's Azadi Tower lights in the colours of the Flag of China to show support On 28 January, the city of Mito donated 50,000 masks to its sister city of Chongqing, and on 6 February, the city of Okayama sent 22,000 masks to Luoyang, its sister city. On 10 February, the ruling Liberal Democratic Party (LDP) of Japan made a symbolic deduction of 5,000 yen from the March salary of every LDP parliamentarian, a total of two million yen, to donate to China; party secretary general Toshihiro Nikai stated that "For Japan, when it sees a virus outbreak in China, it is like seeing a relative or neighbour suffering. Japanese people are willing to help China and hope the outbreak will pass as soon as possible."

Other countries have also announced aid efforts. Malaysia announced a donation of 18 million medical gloves to China. The Philippine Red Cross donated $1.4 million worth of Philippine-made face masks to Wuhan. Turkey dispatched medical equipment, and Germany delivered various medical supplies including 10,000 Hazmat suits. On 19 February, the Singapore Red Cross announced that it would send $2.26 million worth of aid to China, consisting of protective material and training. Russia sent more than 23 tonnes of humanitarian aid to Wuhan.

In March 2020, China, Cuba and Russia sent medical supplies and experts to help Italy deal with its coronavirus outbreak. On 22 March 2020, after a phone call with Italian Prime Minister Giuseppe Conte, Russian president Vladimir Putin arranged the Russian army to send military medics, special disinfection vehicles and other medical equipment to Italy. Businessman Jack Ma sent 1.1 million testing kits, 6 million face masks and 60,000 protective suits to Addis Ababa for distribution by the African Union, as concern grows that poor health infrastructure and high levels of HIV in the region could precipitate severe disruption.

WHO response measures

The WHO has commended the efforts of Chinese authorities in managing and containing the epidemic, with Director-General Tedros Adhanom expressing "confidence in China's approach to controlling the epidemic" and calling for the public to "remain calm". The WHO noted the contrast between the 2003 epidemic, where Chinese authorities were accused of secrecy that impeded prevention and containment efforts, and the current crisis where the central government "has provided regular updates to avoid panic ahead of Lunar New Year holidays".

WHO director-general Tedros Adhanom

On 23 January, in reaction to the central authorities' decision to implement a transportation ban in Wuhan, WHO representative Gauden Galea remarked that while it was "certainly not a recommendation the WHO has made", it was also "a very important indication of the commitment to contain the epidemic in the place where it is most concentrated" and called it "unprecedented in public health history".

On 30 January, following confirmation of human-to-human transmission outside China and the increase in the number of cases in other countries, the WHO declared the outbreak a Public Health Emergency of International Concern (PHEIC), the sixth PHEIC since the measure was first invoked during the 2009 swine flu pandemic. Tedros clarified that the PHEIC, in this case, was "not a vote of no confidence in China", but because of the risk of global spread, especially to low- and middle-income countries without robust health systems. In response to the implementations of travel restrictions, Tedros stated that "there is no reason for measures that unnecessarily interfere with international travel and trade" and that "WHO doesn't recommend limiting trade and movement."

On 5 February, the WHO appealed to the global community for a $675 million contribution to fund strategic preparedness in low-income countries, citing the urgency to support those countries which "do not have the systems in place to detect people who have contracted the virus, even if it were to emerge". Tedros further made statements declaring that "We are only as strong as our weakest link" and urged the international community to "invest today or pay more later".

On 11 February, the WHO in a press conference established COVID-19 as the name of the disease. On the same day, Tedros stated that UN Secretary-General António Guterres had agreed to provide the "power of the entire UN system in the response". A UN Crisis Management Team was activated as a result, allowing co-ordination of the entire United Nations response, which the WHO states will allow them to "focus on the health response while the other agencies can bring their expertise to bear on the wider social, economic and developmental implications of the outbreak".

WHO representatives holding joint meeting with Tehran administrators

On 14 February, a WHO-led Joint Mission Team with China was activated to provide international and WHO experts to touch ground in China to assist in the domestic management and evaluate "the severity and the transmissibility of the disease" by hosting workshops and meetings with key national-level institutions to conduct field visits to assess the "impact of response activities at provincial and county levels, including urban and rural settings".

On 25 February, the WHO declared that "the world should do more to prepare for a possible coronavirus pandemic," stating that while it was still too early to call it a pandemic, countries should nonetheless be "in a phase of preparedness". In response to a developing outbreak in Iran, the WHO sent a Joint Mission Team there on the same day to assess the situation.

On 28 February, WHO officials said that the coronavirus threat assessment at the global level would be raised from "high" to "very high", its highest level of alert and risk assessment. Mike Ryan, executive director of WHO's health emergencies program, warned in a statement that "This is a reality check for every government on the planet: Wake up. Get ready. This virus may be on its way and you need to be ready," urging that the right response measures could help the world avoid "the worst of it". Ryan further stated that the current data did not warrant public health officials to declare a global pandemic, saying that such a declaration would mean "we're essentially accepting that every human on the planet will be exposed to that virus."

On 11 March, the WHO declared the coronavirus outbreak an official pandemic. The Director-General said that WHO was "deeply concerned both by the alarming levels of spread and severity, and by the alarming levels of inaction".

****

Chapter – 8

Socio-Political consequences of COVID-19

Politics

A number of provincial-level administrators of the Communist Party of China (CPC) were dismissed over their handling of the quarantine efforts in Central China, a sign of discontent with the political establishment's response to the outbreak in those regions. Some experts believe this is likely in a move to protect Communist Party general secretary Xi Jinping from people's anger over the coronavirus outbreak. The Italian government has criticised the European Union's lack of solidarity with coronavirus-affected Italy.

The outbreak has prompted calls for the United States to adopt social policies common in other wealthy countries, including universal health care, universal child care, paid family leave, and higher levels of funding for public health. Political analysts anticipated it may negatively affect Donald Trump's chances of re-election in the 2020 presidential election.

The Government of the Islamic Republic of Iran has been heavily affected by the virus. Iran's President Hassan Rouhani wrote a public letter to world leaders asking for help, saying that his country doesn't have access to international markets due to the United States sanctions against Iran. Diplomatic relations between Japan and South Korea worsened due to the 2020 coronavirus pandemic. South Korea criticised Japan's "ambiguous and passive quarantine efforts".

Education

Learners affected by school closures caused by COVID-19

- Localised school closures
- Country-wide school closures
- No school closures

As of 20 March, more than 960 million children and other students were affected by temporary or indefinite government-mandated school closures. Of these, 105 countries shut schools nationwide, affecting students who would normally attend pre-primary to upper-secondary classes, and 15 countries implemented localized closures, affecting an additional 640 million school children and other students.

On 23 March 2020, Cambridge International Examinations (CIE) released a statement announcing the cancellation of Cambridge IGCSE, Cambridge O Level, Cambridge International AS & A Level, Cambridge AICE Diploma and Cambridge Pre-U examinations for the May/June 2020 series across all countries.

Even when school closures were temporary, the measures carried high social and economic costs, affecting people across communities, but their impact was more severe for disadvantaged children and their families, causing interrupted learning, compromised nutrition, childcare problems and consequent economic cost to families who could not work.

In response to school closures, UNESCO recommended the use of distance learning programs, open educational applications and platforms that schools and teachers can use to reach learners remotely and limit the disruption of education.

Socioeconomics:

Coronavirus fears have led to panic buying of essentials in Singapore and elsewhere, including toilet paper, dried and/or instant noodles, bread, rice, vegetables, disinfectant and rubbing alcohol

The coronavirus outbreak has been attributed to several instances of supply shortages, stemming from: globally increased usage of equipment to fight the outbreaks, panic buying and disruption to factory and logistic operations. The United States Food and Drug Administration has issued warnings about shortages to drugs and medical equipment due to increased consumer demand and supplier disruption. Several localities, such as the United States, Italy, and Hong Kong, also witnessed panic buying that led to shelves being cleared of grocery essentials such as food, toilet paper and bottled water, inducing supply shortages. The technology industry in particular has been warning about delays to shipments of electronic goods. According to WHO director-general Tedros Adhanom, the demand for personal protection equipment has risen 100-fold and this demand has led to the increase in prices of up to twenty times the normal price and also induced delays on the supply of medical items for four to six months. This has also caused a shortage of personal protective equipment worldwide, with the WHO warning that this will endanger health workers.

In Australia, the pandemic has provided a new opportunity for daigou shoppers to sell Australian product into China. This activity has left locals without essential supplies.

As mainland China is a major economy and a manufacturing hub, the viral outbreak has been seen to pose a major destabilizing threat to the global economy. Agathe Demarais of the Economist Intelligence Unit has forecast that markets will remain volatile until a clearer image emerges on potential outcomes. In January 2020, some analysts estimated that the economic fallout of the epidemic on global growth could surpass that of the SARS outbreak. One estimate from an expert at Washington University in St. Louis gave a $300+ billion impact on the world's supply chain that could last up to two years. Organization of the Petroleum Exporting Countries reportedly "scrambled" after a steep decline in oil prices due to lower demand from China. Global stock markets fell on 24 February due to a significant rise in the number of COVID-19 cases outside mainland China. On 27 February, due to mounting worries about the coronavirus outbreak, various US stock indexes including the NASDAQ-100, the S&P 500 Index and the Dow Jones Industrial Average, posted their sharpest falls since 2008, with the Dow falling 1,191 points, the largest one-day drop since the financial crisis of 2007–08. All three indexes ended the week down more than 10%. On 28 February, Scope Ratings GmbH affirmed China's sovereign credit rating, but maintained a Negative Outlook. Stocks plunged again based on coronavirus fears, the largest fall being on 16 March 2020. Many consider an economic recession to be likely.

The near-empty arrival hall of Seoul–Incheon International Airport in South Korea on 6 March

Tourism is one of the worst affected sectors due to travel bans, closing of public places including travel attractions, and advise of governments against any travel all over the world. As a consequence, numerous airlines have cancelled flights due to lower demand, including British Airways, China Eastern and Qantas, while British regional airline Flybe collapsed. Several train stations and ferry ports have also been closed. The epidemic coincided with the Chunyun, a major travel season associated with the Chinese New Year holiday. A number of events involving large crowds were cancelled by national and regional governments, including annual New Year festivals, with private companies also independently closing their shops and tourist attractions such as Hong Kong Disneyland and Shanghai Disneyland. Many

Lunar New Year events and tourist attractions have been closed to prevent mass gatherings, including the Forbidden City in Beijing and traditional temple fairs. In 24 of China's 31 provinces, municipalities and regions, authorities extended the New Year's holiday to 10 February, instructing most workplaces not to re-open until that date. These regions represented 80% of the country's GDP and 90% of exports. Hong Kong raised its infectious disease response level to the highest and declared an emergency, closing schools until March and cancelling its New Year celebrations.

Retail sector has been impacted globally, with store hours reductions or temporary closures. This resulted in 30% drop in daily footfall by 18 March, with additional restrictions, such as closure of all 150+ shopping centres nationally by Simon Property Group, by mall operators around the world.

Despite the high prevalence of COVID-19 cases in Northern Italy and the Wuhan region, and the ensuing high demand for food products, both areas have been spared from acute food shortages. Effective measures by China and Italy against the hoarding and illicit trade of critical products have been carried out with success, avoiding acute food shortages that were anticipated in Europe as well as in North America. Northern Italy with its significant agricultural production has not seen a large reduction, but prices may increase according to industry representatives. Empty food shelves were only encountered temporarily, even in Wuhan city, while Chinese government officials released pork reserves to assure sufficient nourishment of the population. Similar laws exist in Italy that require food producers to keep reserves for such emergencies.

Environment

Empty motorway in the Netherlands.

Due to the coronavirus outbreak's impact on travel and industry, many regions experienced a drop in air pollution. The Centre for Research on Energy and Clean Air reported that methods to contain the spread of coronavirus, such as quarantines and travel bans, resulted in a 25% reduction of carbon emission in China. In the first month of lockdowns, China produced approximately 200 million fewer metric tons of carbon dioxide than the same period in 2019, due to the reduction in air traffic, oil refining, and coal consumption. Between 1 January and 11 March 2020, the European

Space Agency observed a marked decline in nitrous oxide emissions from cars, power plants and factories in the Po Valley region in northern Italy, coinciding with lockdowns in the region. In Venice, the water in the canals cleared up and experienced an increased presence of fish and waterfowl; the Venice mayor's office clarified that the increase in water clarity was due to the settling of sediment that is disturbed by boat traffic and mentioned the decrease in air pollution along the waterways.

Despite a temporary decline in global carbon emissions, the International Energy Agency warned that the economic turmoil caused by the coronavirus outbreak may prevent or delay companies from investing in green energy. However, extended quarantine periods have boosted adoption of remote work policies.

As a consequence of the unprecedented use of disposable face masks, significant numbers are entering the natural environment and in particular, to rivers and seawater. In some cases, the masks have been washed onto beaches where they are accumulating. This accumulation has been reported on beaches in Hong Kong and is expected to add to the worldwide burden of plastic waste and the detrimental effects of this waste to marine life.

Culture

The closed entrance to the Shah Abdol-Azim Shrine in Ray, Iran. Cultural sites and other public spaces throughout the world were closed to slow the spread of the pandemic.

Another recent and rapidly accelerating fallout of the disease is the cancellation of religious services, major events in sports, the film industry, and other societal events, such as music festivals and concerts, technology conferences, fashion shows and sports.

The Vatican announced that Holy Week observances in Rome, which occur during the last week of the Christian penitential season of Lent, have been cancelled. Many dioceses have recommended older Christians to stay at home rather than attending Mass on Sundays; some churches have made church services available via radio, online livestreaming or television while others are offering drive-in worship. With the Roman Catholic Diocese of Rome closing its churches and chapels and St. Peter's Square is emptied of Christian pilgrims, other religious bodies also cancelled services and limiting public gatherings in churches, mosques, synagogues and gurdwaras. Iran's

Health Ministry announced the cancellation of Friday prayers in areas affected by the outbreak and shrines were later closed, while Saudi Arabia banned the entry of foreign pilgrims as well as its residents to the holy sites in Mecca and Medina.

The pandemic has caused the most significant disruption to the worldwide sporting calendar since the Second World War. Most major sporting events scheduled to take place now or in the next few weeks have been either cancelled or postponed, including the 2019–20 UEFA Champions League, 2019–20 Premier League, UEFA Euro 2020, 2019–20 NBA season, and 2019–20 NHL season. The outbreak disrupted plans for the 2020 Summer Olympics (originally scheduled to start at the end of July); the International Olympic Committee announced on 24 March that the event will be "rescheduled to a date beyond 2020 but not later than summer 2021".

While casinos and other gaming venues worldwide have been closed and live poker tournaments have been either postponed or cancelled, gamblers moved online with many online gambling sites reporting doubling of their rate of new sign-ups.

The entertainment industry has also been affected, with various music groups suspending or cancelling concert tours. Many large theatres such as those on Broadway also suspended all performances. Some artists have explored ways to continue to produce and share work over the internet as an alternative to traditional live performance, such as live streaming concerts or creating web-based "festivals" for artists to perform, distribute and publicize their work.

Potential long-term impacts

The political, cultural, and socio-economic impacts of the pandemic may together cause major changes in human society. Commentators have suggested this could include an increase in remote work, localization of global supply chains, and increased political polarization. India imposed a 21-day lockdown in March 2020, and Trump called for the United States to be open "by Easter."

It has been suggested that the pandemic could cause a "revolution in working from home" as many major corporations have offered telecommuting in order to slow the spread of the virus. This effect has also been observed in the rise of online learning in higher education. Additionally, there could also be a longer-term reduction in business travel and international conferences, with virtual substitutes replacing them. A more widespread reversal of globalization has also been discussed, especially in regard to supply chains. The German Economy Minister has supported localization of supply chains in reaction to the pandemic.

Xenophobia and racism

Houston Chinatown where businesses claim up to a 75% reduction in business due to the pandemic.

Since the outbreak of COVID-19, heightened prejudice, xenophobia and racism have been noted toward people of European, Chinese and other East Asian descent, as incidents of fear, suspicion and hostility have been observed in many countries, particularly in Europe, East Asia, North America and the Asia-Pacific region. Some countries in Africa saw rising anti-Chinese sentiment as well. There has been support for the Chinese, both on and offline, towards those in virus-stricken areas, but many residents of Wuhan and Hubei have reported experiencing discrimination based on their regional origin. Since the progression of the outbreak to new hot-spot countries, people from Italy, the centre of Europe's coronavirus outbreak, have also been subjected to suspicion and xenophobia. Islamists have exploited the disease to foster anti-Western sentiment. In Hong Kong, anti-Western sentiment has risen as expats are accused of introducing a 'second wave' of the disease. Anti-Caucasian sentiment has also risen in Thailand, where Westerners are accused of spreading the disease.

Citizens in countries including Malaysia, New Zealand, Singapore, and South Korea, initially signed petitions lobbying their government to ban Chinese people from entering their countries in an effort to stop the disease spreading. In Japan, the hashtag #ChineseDontComeToJapan had been trending on Twitter. Chinese people in the United Kingdom said they were facing increasing levels of racist abuse, with cases of assaults reported. Protesters in Ukraine attacked buses carrying Ukrainian and foreign evacuees from Wuhan. Students from Northeast India, which shares a border with China, who study in major Indian cities have reportedly experienced harassment related to the coronavirus outbreak. The Bharatiya Janata Party's State unit president in West Bengal Dilip Ghosh stated that the Chinese had destroyed nature and "that's why the God took revenge against them." The remarks were later condemned by the Chinese consulate in Kolkata, calling it "erroneous".

There were rumors in India that some Muslims who returned from Dubai refused to undergo coronavirus testing for religious reasons. The news was then amplified by bloggers and some some social media channels, but it was later debunked and confirmed to be false. Local authorities in Bolivia quarantined Japanese nationals despite them having no coronavirus-related symptoms. In the Russian cities of Moscow and Yekaterinburg, Chinese nationals were targeted by quarantine enforcing campaigns, as well as police raids, which were condemned by human rights advocates as racial profiling. The Chinese Embassy in Germany has acknowledged a rise in hostile cases against its citizens since the outbreak. Children of Asian descent were ostracized and mocked over their origins in middle schools near Paris. Many French-Vietnamese report also being subject to harassment since the coronavirus outbreak in Wuhan.

As the pandemic has progressed, there have also been isolated instances of prejudice against Westerners, particularly Western tourists accused of importing the disease. Most notably, a post to the Twitter account of the Thai Health Minister, Anutin Charnvirakul, urged Thais to be "more careful of Westerners than Asians". It claimed many Western tourists refused to wear masks and "dressed dirtily and never shower". The post attracted negative feedback in Thailand and the UK and was quickly removed. Mr Charnvirakulk claimed he was not responsible for it.

On 30 January, the WHO's Emergency Committee issued a statement advising all countries to be mindful of the "principles of Article 3 of the IHR (the International Health Regulations)", which the WHO says is a caution against "actions that promote stigma or discrimination" when conducting national response measures to the outbreak.

****

Chapter – 9

Case Study-i

Climate and Socio-economic impact of COVID-19: A Case Study in India

Kousik Das Malakar[1]

Abstract:

Aims and Objectives: The present study focuses on determining the status of Climate and Socio-economic impact of COVID-19 in India. The study seeks to examine that, (a) the changing situation of climatic phenomena (air pollutants) and determining the Air Quality Index (AQI), (b) Define the impacts of various socio-economic sectors due to COVID-19 in India, and (c) estimate the economic impact from COVID-19 on India's GVA April-June 2020 by sector.

Data and Methods: Data set compiled from Johns Hopkins University, USA; Online COVID-19, Govt. of India; Central Pollution Control Board, Govt. of India, Statista 2020. And using the MS Excel 2014 & ArcGIS 10.2.2 applications.

Results: (a) AQI has been decreasing and showing satisfactory to good quality of air pollutant conditions. (b) Due to COVID-19, India's all socio-economic sectors have running negative-way (Presently). (This studies based on the month of January to 26th of April).

Keywords: COVID-19; Socio-economic Impacts; Climatic Impacts; India

----

[1]Centre for the Study of Regional Development (CSRD), School of Social Sciences (SSS), Jawaharlal Nehru University (JNU), New Delhi, India. E-mail: kdmalakargegr92@gmail.com

1. Introduction:

COVID-19 is a pneumonia-like disease with symptoms including fever, dry cough, Tightness of Chest, Head Ache, shortness of breath and Kidney Failure that was first identified in Wuhan, China (American society for Microbiology, 2020 and WHO, 2020). Mainly, older adults and People who have serious chronic medical conditions like, Heart disease Diabetes Lung disease suppressed immune systems, whose are more affected than the others (CDS, 2020). COVID-19 is transmitted human to human by the medium of air by coughing and sneezing, Close personal contact, such as touching or shaking hand, touching an object or surface with the virus on it, then touching your mouth, nose, or eyes before washing your hands (Malakar, 2020).

In India, the first case of COVID-19 was reported on January 30, 2020. As of April 26, confirmed cases of COVID-19 stand at 26,377. While 19,618 patients have active cases, 5,938 have recovered, and 826 have died (Govt. of India, 26th April, 2020). India has taken early action to limit the spread of COVID-19, ordering a 21-day nationwide lockdown for its population of 1.3 billion people starting March 25. The novel coronavirus has spread widely in India relatively recently compared to other countries, and the number of reported infections is low so far, with 5,274 cases and 149 deaths as of April 8. However, as COVID-19 cases are increasing fast, there's great concern about the disease's potential spread and impact. India has to be ready for a possible surge. Testing should be expanded significantly. The government views the pattern of the spread of COVID-19 as almost like the 2009 H1N1 influenza pandemic, meaning the spread is unlikely to be uniform. After the 21-day period expires, it is planning to maintain the full lockdown in "hotspot" areas and relax it in other places (Dev, 2020).

These measures may help in limiting the health crisis, but as in other countries, the complete shutdown of all economic activities except essential services will create an economic crisis and misery for the poor, with massive job losses and rising food insecurity.

The economic shock will likely be much more severe for India, for two reasons. First, pre-COVID-19, the economy was already slowing down, compounding existing problems of unemployment, low incomes, rural distress, malnutrition, and widespread inequality. Second, India's large informal sector is particularly vulnerable. Out of the national total of 465 million workers, around 92.8% were informal workers in 2017-18 (Census of India, 2018). Lacking regular salaries or incomes, these agriculture, migrant, and other informal workers would be hardest-hit during the lockdown period. Mainly, its fall the effects on various socio-economic activities (The Ministry of Labour & Employment, Govt. of India, 2020) as well as climatic phenomenon (Central Pollution Control Board, Govt. of India, 2020) also.

In this paper focuses the main problems of COVID-19 in India according to the date of 26th April, 2020 (perspective of climate and socio-economic). Here, focus on the impacts on climatic pollutants (PM 2.5, PM 10, N2O, NH3, and O3), where on its represent the Air Quality Index (AQI), and Socio-economic activities (Agriculture, supply chains, food, and nutrition security, livelihoods, Economic Impacts, Manufacturing, Defence, Migrant workers and labour force, Suspension of MPLADS).

2. Objectives:

Therefore, the specific objectives of this study are;

1) To evaluate the changing situation of climatic phenomena (air pollutants).
2) Define the impacts of various socio-economic sectors due to COVID-19 in India.
3) To estimate the economic impact from COVID-19 on India's GVA April-June 2020 by sector.

3. Study Area:

The study area is India. Mainly focuses on the two fields, there are such as, (a) 23 cities (Map 1) for evaluating the changing situation of climatic phenomena (air pollutants) due to COVID-19. (b) State-wise (Map 2) assessment of the impacts of various socio-economic sectors.

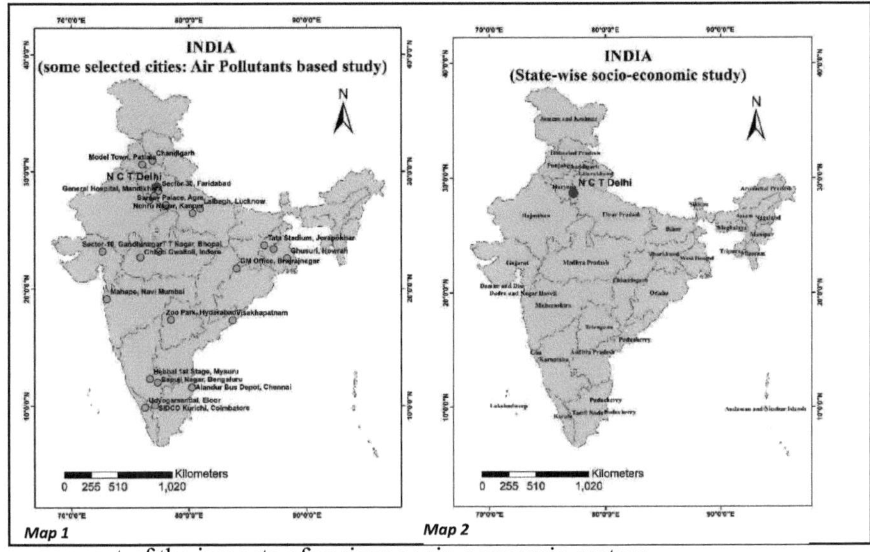

Map 1 *Map 2*

4. Database and Methodology:

For the present study data set compiled from various secondary sources like, [2]Worldwide COVID-19 data from Johns Hopkins University, USA; [3]Indian state-wise COVID-19 data from Online COVID-19, Govt. of India; [4]Air pollutants data from Central Pollution Control Board, Govt. of India, [5]Estimated impact from the coronavirus (COVID-19) on India between April and June 2020, by sector GVA data from Statista 2020.

[2]https://www.channelnewsasia.com/
[3]https://www.covid19india.org/
[4]https://app.cpcbccr.com/AQI_India/
[5]https://www.statista.com/

Also calculate the data by using MS Excel format 2014 and ArcGIS 10.2.2 and prepared the mapping used the various types of techniques of ArcGIS application 10.2.2 and figurate the bar diagram using the MS Excel format 2014.

5. About COVID-19:

Coronaviruses constitute the subfamily Orthocoronavirinae, within the family Coronaviridae, order Nidovirales, and realm Riboviria. They have enveloped viruses with a positive-sense single-stranded RNA genome and a nucleocapsid of helical symmetry. The genome size of coronaviruses ranges from approximately 27 to 34 kilobases, the most important among known RNA viruses. The name coronavirus springs from the Latin corona, meaning "crown" or "halo", which refers to the characteristic appearance like a crown or a solar corona around the virions (virus particles) when viewed under two-dimensional transmission microscopy, thanks to the surface being covered in club-shaped protein spikes.

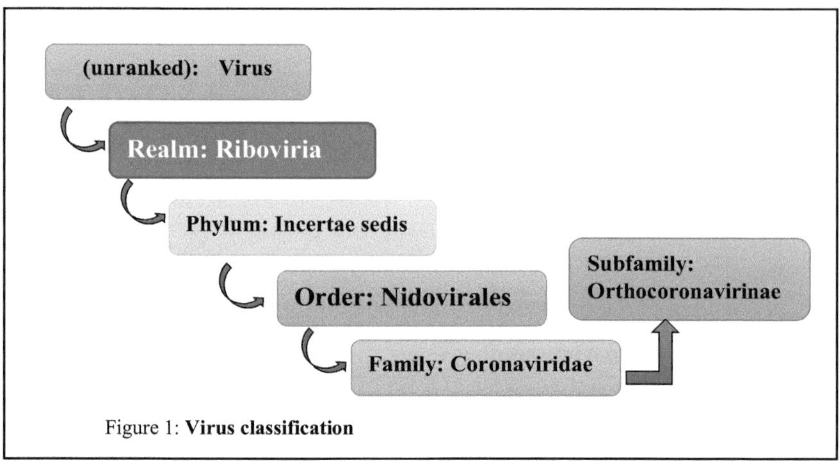

Figure 1: **Virus classification**

Various Names:

Firstly, COVID-19 also known as, Coronavirus disease 2019, Novel coronavirus pneumonia, 2019-nCoV acute respiratory disease, and Wuhan coronavirus also.

Symptoms:

COVID-19 symptoms range from mild to severe. It takes 2-14 days after exposure for symptoms to develop. Symptoms may include:

• Fever • Difficulty in Breathing • Coughing • Tightness of Chest • Running Nose • Head Ache • Feeling of being Unwell • Pneumonia • Kidney Failure.

Risk Persons:

According to the CDC, children don't seem to be at higher risk for getting COVID-19. However, some people are, including Older adults and People who have serious chronic medical conditions like: Heart disease Diabetes Lung disease Suppressed immune systems.

Causes a Coronavirus Infection:

Humans rst get a coronavirus from contact with animals. Then, it can spread from human to human. Health ocials do not know what animal caused COVID-19.

The COVID-19 virus can be spread through contact with certain bodily uids, such as droplets in a cough. It might even be caused by touching something an infected person has touched then touching your hand to your mouth, nose, or eyes.

Mode of transmission:

Human Coronavirus (2019-nCoV) most ordinarily spread from an infected person to other through:

- The air by coughing and sneezing.
- Close personal contact, such as touching or shaking hand.
- Touching an object or surface with the virus on it, then touching your mouth, nose, or eyes before washing your hands.

5. Results and Discussion:

5.1 Present Situation:

5.1.1 Worldwide:

As per the date of 26th April 2020, worldwide COVID-19 has fall the very horror effects on USA more than 9 lacks peoples are affected, Spain has more than 2 lacks and the position of India has affected in COVID-19 more

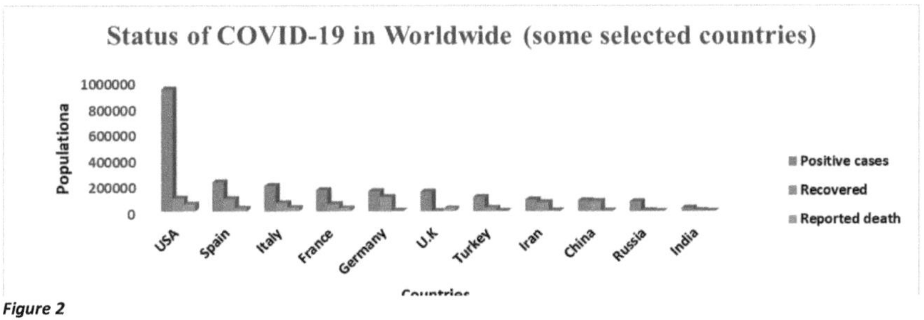

Figure 2

than 26, 000. Following figure 2 shows the status of some selected countries COVID-19's situations. (Source: Johns Hopkins University, USA (26th April 2020)).

5.1.2 India:

As per the date of 26th April 2020, India (state-wise) affected due to COVID-19 are more than 26, 000. Most of vulnerable state is Maharashtra (More than 7000), than Gujarat (More than 3000), Delhi (More than 2500), Rajasthan (More than 2000), Madhya Pradesh (More than 1900), Tamil Nadu (More than 1800) and others. Following figure 3 shows the status of state-wise COVID-19's situations. (Source: COVID-19, Govt. of India (https://www.covid19india.org/))

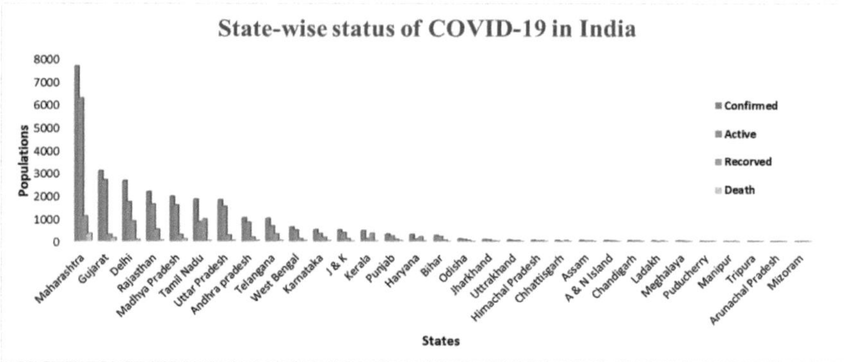

Figure 3

5.2 Impacts on Climate:

Coronavirus (COVID-19) has fall the effect of Indian Climate. In this studies, shows that types of findings. Here, 23 cities have been selected for the determining of AQI and their respective pollutants (PM 2.5, PM 10, N2O, NH3, O3).

5.2.1 Distribution of PM 2.5 and PM 10:

The size of particles affects their potential to cause health problems:

PM10 (particles with a diameter of 10 micrometres or less): these particles are sufficiently small to undergo the throat and nose and enter the lungs. Once inhaled, these particles can affect the guts and lungs and cause serious health effects.

PM2.5 (particles with a diameter of two .5 micrometres or less): these particles are so small they will get deep into the lungs and into the bloodstream. There is sufficient evidence that exposure to PM2.5 over long

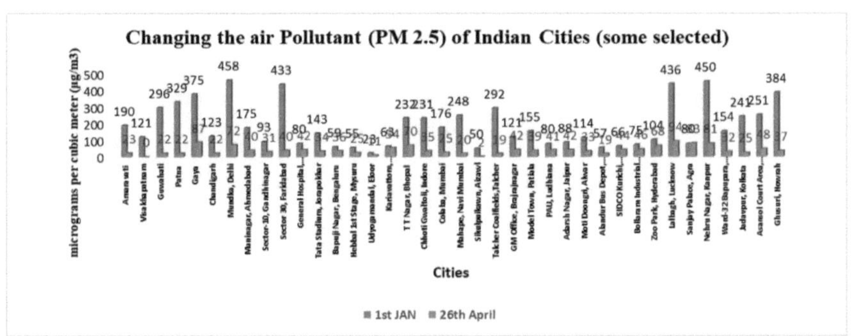

Figure 5

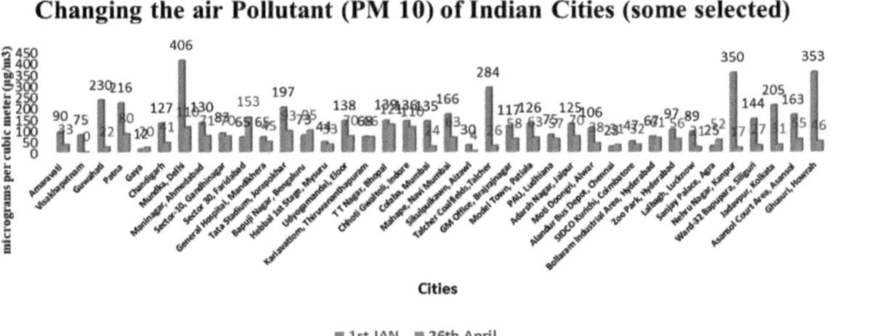

Figure 4

periods (years) can cause adverse health effects. Note that PM10 includes PM2.5.

Figure 4 and 5 shows the changing air pollutant conditions of Indian cities due to lack of transportation and others in this lockdown period. Here, shows that, the changeness between pre-lockdown time and present (26th April) lockdown period. Air pollutants has been decrease.

5.2.2 Distribution of N2O:

Nitrogen dioxide may be a highly reactive gas formed by emissions from automobiles, industry, unflued gas-heaters and gas range tops. High concentrations are often found especially near busy roads and indoors where unflued gas-heaters are in use.

Potential health effects from exposure to nitrogen dioxide:

- Increased susceptibility to lung infections in people with asthma.
- Increased susceptibility to asthma triggers like pollen and exercise.
- Worsened symptoms of asthma – more frequent asthma attacks.
- Airway inflammation in healthy people.

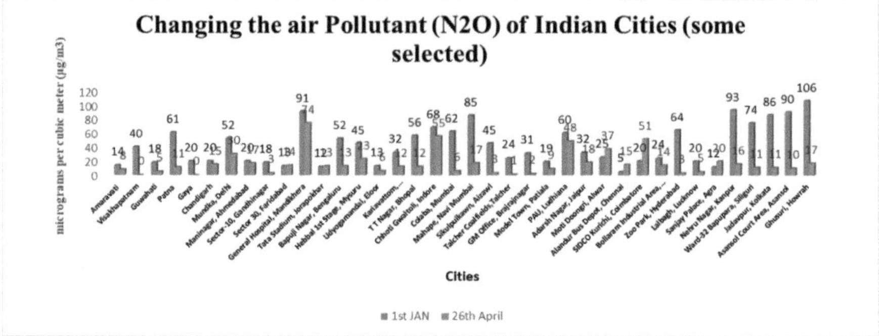

Figure 6

5.2.3 Distribution of O3:

Ozone, O3, consists of three oxygen atoms joined together. Two oxygen atoms joined together form the essential oxygen molecule O2. the extra third atom makes ozone an unstable, highly reactive gas. Ozone is found in two areas of the Earth's atmosphere: within the upper atmosphere and at ground level. Ozone within the upper atmosphere protects us by filtering out damaging ultraviolet from the sun.

Potential health effects from exposure to ozone:

- Irritation and inflammation of eyes, nose, throat and lower airways: coughing, sore and scratchy throat or uncomfortable feeling in chest.
- Reduced lung function: unable to breathe as deeply or vigorously as you normally would.
- Exacerbation of asthma and chronic respiratory diseases like bronchitis (also called chronic obstructive pulmonary disease or COPD).
- Increased susceptibility to respiratory infections.

- Can still damage lungs when symptoms have disappeared.

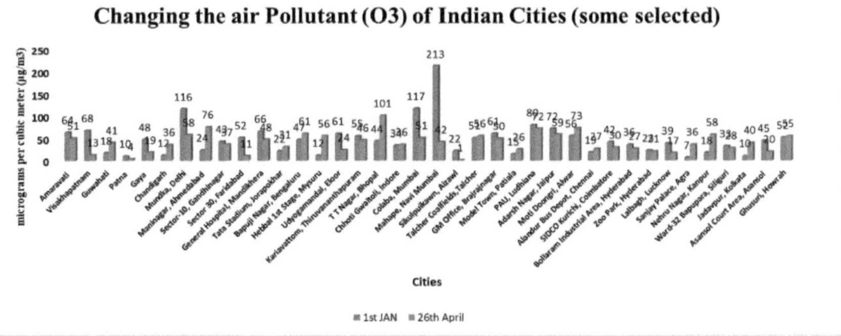

Figure 7

Figure 6 and 7 shows the changing air pollutant conditions of Indian cities due to lack of transportation and others in this lockdown period. Here, shows that, the changeness between pre-lockdown time and present (26th April) lockdown period. Air pollutants has been decrease. And decrease the air borne diseases also and environment play freely.

5.2.4 Air Quality Index:

An air quality index (AQI) is employed by government agencies to speak to the general public how polluted the air currently is or how polluted it's forecast to become. Public health risks increase because the AQI rises. Following table 1. shows that the AQI and their respective remarks.

Table 1.

AQI	Remarks
0 – 50	Good
51-100	Satisfactory
101-200	Moderate
201-300	Poor
301-400	Very Poor
401-500	Severe
Source: Central Pollution Control Board, Govt. of India	

Following Map 3 and 4 show the AQI of 23 selected cities of India. and represent the very good conditions for the Environment. That is possible due to the COVID-19 lockdown situation.

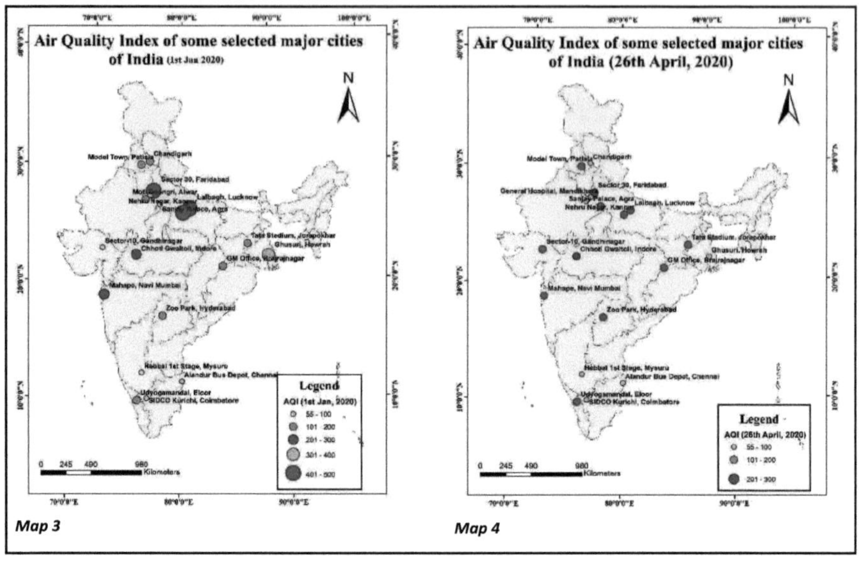

Map 3

Map 4

Status of the AQI is given below table 3.

Table 3. Air Quality Index of Indian Major cities					
States/ Uts	**Cities**	**1st January**	**Remarks**	**26th April**	**Remarks**
AP	Secretariat, Amaravati	190	Moderate	33	Good
	UVM Visakhapatnam	121	Moderate	0	Good
Assam	Guwahati	296	Poor	22	Good
Bihar	Patna	329	Very Poor	80	Satisfactory
	Gaya	375	Very Poor	87	Satisfactory
Chandigarh	Chandigarh	127	Moderate	41	Good
Delhi	Mundka, Delhi	458	Severe	110	Moderate
Gujarat	Maninagar, Ahmedabad	175	Moderate	71	Satisfactory
	Sector-10, Gandhinagar	93	Satisfactory	70	Satisfactory

Haryana	Sector 30, Faridabad	433	Severe	153	Moderate
	General Hospital, Mandikhera	80	Satisfactory	45	Satisfactory
Jharkhand	Tata Stadium, Jorapokhar	197	Moderate	93	Satisfactory
Karnataka	Bapuji Nagar, Bengaluru	59	Satisfactory	95	Satisfactory
	Hebbal 1st Stage, Mysuru	55	Satisfactory	33	Good
Kerala	Udyogamandal, Eloor	138	Moderate	70	Satisfactory
	Kariavattom, Thiruvananthapuram	63	Satisfactory	66	Satisfactory
Madhya Pradesh	T T Nagar, Bhopal	232	Poor	121	Moderate
	Chhoti Gwaltoli, Indore	231	Poor	110	Moderate
Maharashtra	Colaba, Mumbai	176	Moderate	25	Good
	Mahape, Navi Mumbai	248	Poor	63	Satisfactory
Mizoram	Sikulpuikawn, Aizawl	50	Good	4	Good
Odisha	Talcher Coalfields,Talcher	292	Poor	26	Good
	GM Office, Brajrajnagar	121	Moderate	58	Satisfactory
Punjab	Model Town, Patiala	155	Moderate	63	Satisfactory
	PAU, Ludhiana	80	Satisfactory	57	Satisfactory
Rajasthan	Adarsh Nagar, Jaipur	125	Moderate	70	Satisfactory
	Moti Doongri, Alwar	114	Moderate	38	Good
Tamil Nadu	Alandur Bus Depot, Chennai	57	Satisfactory	31	Good
	SIDCO Kurichi, Coimbatore	66	Satisfactory	44	Good
Telangana	Bollaram Industrial Area, Hyderabad	75	Satisfactory	61	Satisfactory
	Zoo Park, Hyderabad	104	Moderate	68	Satisfactory

U. P	Lalbagh, Lucknow	436	Severe	94	Satisfactory
	Sanjay Palace, Agra	80	Satisfactory	83	Satisfactory
	Nehru Nagar, Kanpur	450	Severe	81	Satisfactory
W.B.	Ward-32 Bapupara, Siliguri	154	Moderate	32	Good
	Jadavpur, Kolkata	241	Poor	31	Good
	Asansol Court Area, Asansol	251	Poor	55	Satisfactory
	Ghusuri, Howrah	384	Very Poor	46	Good
Source: Central Pollution Control Board, Govt. of India					

5.3 Impacts on socio-economic field:

5.3.1 Agriculture and provide chains:

COVID-19 is disrupting some activities in agriculture and provide chains. Preliminary reports show that the non-availability of migrant labour is interrupting some harvesting activities, particularly in northwest India where wheat and pulses are being harvested. There are disruptions in supply chains due to transportation problems and other issues. Prices have declined for vegetables, wheat, and other crops, yet consumers are often paying more. Media reports show that the closure of restaurants, hotels, other shops, and tea shops during the lockdown is already depressing milk sales. Meanwhile, poultry farmers are badly hit thanks to misinformation, particularly on social media, that chicken is the carriers of COVID-19.

5.3.2 Sector Wise economic Impact:

5.3.2.1 Agriculture:

The nationwide lockdown has left farmers across the country bereft of agricultural labour just before the crucial harvesting season. Farmers also worry about government procurement and their ability to sell their crops, given that many agricultural markets are still closed, despite orders from the home ministry to exempt all farming activities from the shutdown. Unless the government acts soon, farmers in India will face a bleak future leading to bankruptcies and farmer suicides. This will severely dent India's ability to revive its economy since there will be a sharp decline in consumption in the rural sectors leading to impairment of the industrial activity.

5.3.2.2 Textile:

Production halts in China and lockdown in India have had an impact. Mainly because of dependence on China for textile raw materials including synthetic yarn, synthetic fabric, buttons, zippers, and hangers. India also exports cotton yarn to China in bulk quantity, and poor demand in China has caused cotton prices to come down in India.

5.3.2.3 Apparel:

Payment of at least US$2 billion is stuck with foreign buyers because of deferment or cancellation of orders. Closure of retail stores across the country is also adding to losses.

5.3.2.4 Automotive:

The sector was already witnessing sluggish demand for the last one year. The present situation has further aggravated the problem and compounded the situation with an acute liquidity crunch. China accounts for 27 percent of India's automotive part imports. With Wuhan being a major auto hub, the supply chain of the automotive sector has been hit significantly.

5.3.2.5 Aviation:

With global travel is suspended, airlines are looking at bankruptcy. It is possible that the government might bail the aviation sector in the near future.

5.3.2.6 Hotels and restaurants:

Demand has declined substantially with owners struggling to recover the fixed costs. Pre-lockdown, hotel occupancy rates of 70 percent declined to 20 percent with restaurants losing 30 to 35 percent of their business.

5.3.2.7 Poultry:

Untrue claims regarding the transmission of COVID-19 through chicken and other meat have impacted the sales and price of poultry items. The sector is facing a loss in the range of US$20 million every day.

5.3.2.8 Chemicals:

Dependence on China for active ingredients and shortage of essential chemicals amidst the lockdown has resulted in reduced production of organic and inorganic chemicals across India.

5.3.2.9 Consumer durables:

Electronic appliances are experiencing a slowdown due to lack of demand, reduced economic activity, and supply chain dependencies with China and other impacted countries. India imports around 50 percent of its completely built units of consumer durables from China.

5.3.2.10 Entertainment and sports:

Places of gathering like cinema halls and malls have been closed. Promotional and sports events are also being called off. One of the major cricketing events in the country – Indian Premier League (IPL) has been deferred for the time being – its cancellation can alone mean a loss of at least US$500 million for the organizing body.

5.3.2.11 FMCG:

After the lockdown announcement, demand for essential FMCG products spiked up owing to hoarding and panic buying by consumers. Grocery items, milk, and hygiene products have seen a surge in demand while supply chain constraints have limited the manufacturing capacities.

5.3.2.12 Pharmaceuticals:

As per the Trade Promotion Council of India, the country imports around 85 percent of its total requirement of active pharmaceutical ingredients (APIs) from China. Dependencies on China have impacted manufacturing operations in India. However, the demand for essential medicines and safety equipment has gone up. The government has restricted the export of certain medicines and pharma products and is monitoring the overall inventory on a regular basis.

5.3.2.13 E-commerce:

Several e-commerce players are unable to service existing orders and are not accepting new orders, even when there is a surge in demand for home delivery. However, companies are trying to service essential items on a priority basis.

5.3.2.14 IT and ITes:

Slowdown in markets has reduced the influx of IT projects from international markets. Remote working has given the rise in demand for communication tools, conference platforms, and cybersecurity apps, among others. These software tools are being used across sectors such as education, finance, and HR to ensure business continuity.

5.3.3 Manufacturing:

Major companies in India such as Larsen and Toubro, Bharat Forge, UltraTech Cement, Grasim Industries, the fashion and retail wing of Aditya Birla Group, Tata Motors and Thermax have temporarily suspended or significantly reduced operations in a number of manufacturing facilities and factories across the country.

5.3.4 Defence:

The Department of Military Affairs led by the Chief of Defence Staff has postponed all capital acquisitions until the coronavirus pandemic recedes. No new major defense deals are going to be made within the beginning of the fiscal year 2020–21.

5.3.5 Migrant workers and labour force:

Due to the lockdown, there suddenly was no work for many daily workers (the urban poor and migrant labourers). At the same time the lockdown restrictions put a stop on the movement of buses and trains. Large numbers of migrant workers ended up walking back to there villages, some journeys hundreds of kilometers long. Commentators commented on how the whole purpose of social distancing was defeated (The Atlantic, 2020; Bloomberg Opinion, 2020).

Soon after a central government directive in late March, state governments set up 21,000 camps to house over 660,000 migrants and stop the exodus (The Tribune, 2020; Politics and Nations, 2020; The Print, 2020). Delhi government is providing free food to 400,000 people a day. Over 500 hunger relief centres are set by the Delhi government (The Times of India 2020).

5.3.6 Suspension of MPLADS:

The Modi government, in view of the coronavirus pandemic, suspended Members of Parliament Local Area Development Scheme (MPLADS) for two years. This action has been called problematic in many ways, including causing a centralisation of power, being anti-federal in nature, and having an affect on local level development and MP influence at micro levels of the society to handle distress. There have been calls for halting the ₹20,000 crore (US$2.8 billion) redevelopment of the central vista project in Delhi instead (The INDIAN Express, 2020: The week, 2020; Business Today, 2020).

5.4 Estimated impact from the coronavirus (COVID-19) on India between April and June 2020, by sector GVA:

Following figure 8 shows the expected impact from the coronavirus (COVID-19) on India between April and June 2020, by sector GVA. All

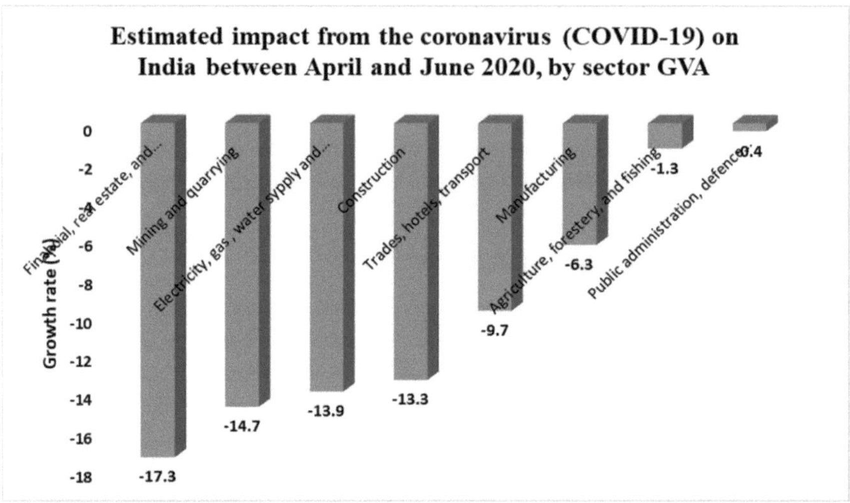

Figure 8 Source: Statista 2020.

sectors shows the negative change for the month of April and June due to lockdown situations and no sufficient mobilization of peoples.

6. Conclusion:

Finally, COVID-19 created a Pandemic situation in the world and its fall effects on Indian climatic and various socio-economic sectors. And it is being created a negative growth rate (figure 8). Presently, fight only COVID-19 and maintain physical distance, not a social distance (it's mandatory to each other contact remotely), and follow the government precautions.

Acknowledgment

The author thank anonymous reviewer for constructive comments. And thanks to Dr. D.N.Das, Dr. Bhaswati Das, Dr. Milap Punia, Dr. Sanjeev Sharma (CSRD, JNU) and Dr. Nieuwaal (Director Delta Alliance, The Netherlands) for the theoretical and applied helps of this paper.

7. References:

1) American society for Microbiology, (2020). https://asm.org/Press-Releases/2020/COVID-19-Resources
2) Bloomberg Opinion (2020). Coronavirus Exposes India's Official Callousness0 (https://www.bloombergquint.com/global-economics/india-s-government-should-have-expected-coronavirus-migration).
3) Business Today (2020). Why suspension of MPLADS is problematic in corona times? (https://www.businesstoday.in/current/economy-politics/coronavirus-in-india-why-suspension-of-mplads-is-problematic/story/400771.html).
4) CDC (https://www.cdc.gov/coronavirus/2019-ncov/specific-groups/high-risk-complications.html)
5) Census of India (https://censusindia.gov.in/)
6) Dev, S.M., (2020). Addressing COVID-19 impacts on agriculture, food security, and livelihoods in India. International Food Policy Research Institute, IFPRI blog pages. (https://www.ifpri.org/blog/addressing-covid-19-impacts-agriculture-food-security-and-livelihoods-india).
7) Govt. of India (COVID-19 database), (26th April, 2020) https://www.covid19india.org/
8) Malakar. K.D., (2020). Coronavirus (COVID-19): A Pandemic Situation of in the World. Lambert Academic Publishing, Germany. ISBN: 978-620-2-51827-7.
9) Politics and Nations (2020). More than 21,000 camps set up for over 6,60,000 migrants: State governments. (https://economictimes.indiatimes.com/news/politics-and-

nation/more-than-21000-camps-set-up-for-over-660000-migrants-state-governments/articleshow/74920798.cms)

10) The Atlantic (2020). The Callousness of India's COVID-19 Response. (https://www.theatlantic.com/international/archive/2020/03/india-coronavirus-covid19-narendra-modi/608896/).

11) The INDIAN Express (2020). Union Government should rethink Central Vista project. (https://www.newindianexpress.com/opinions/2020/apr/25/union-government-should-rethink-central-vista-project-2134979.html).

12) The Print (2020). Rotis, mobile recharges, carrom boards — how Kerala fixed its migrant worker anger. (https://theprint.in/india/rotis-mobile-recharges-carrom-boards-how-kerala-fixed-its-migrant-worker-anger/403937/)

13) The Times of India (2020). Delhi lockdown: Over 500 hunger relief centres set up for 4 lakh people. (https://timesofindia.indiatimes.com/city/delhi/govt-to-give-2-meals-a-day-to-4-lakh-people-says-kejriwal/articleshow/74855058.cms).

14) The Tribune (2020). Over 75 lakh being fed at food camps: MHA. (https://www.tribuneindia.com/news/nation/over-75-lakh-being-fed-at-food-camps-mha-66223).

15) The week (2020). COVID-19: Opposition fumes over Modi govt decision to suspend MPLADS funds. (https://www.theweek.in/news/india/2020/04/06/opposition-parties-hit-out-at-govt-over-mplads-decision-say-will-hit-development-works.html).

16) WHO (https://www.who.int/emergencies/diseases/novel-coronavirus-2019)

The End

Chapter – 10

Case Study-ii

COVID-19 make 2020, a year of uncertainty for the World Economy

COVID-19 is a pneumonia-like disease with symptoms including fever, dry cough, Tightness of Chest, Head Ache, shortness of breath and Kidney Failure that was first identified in Wuhan, China (American society for Microbiology, 2020 and WHO, 2020). Mainly, older adults and People who have serious chronic medical conditions like, Heart disease Diabetes Lung disease suppressed immune systems, whose are more affected than the others (CDS, 2020). COVID-19 is transmitted human to human by the medium of air by coughing and sneezing, Close personal contact, such as touching or shaking hand, touching an object or surface with the virus on it, then touching your mouth, nose, or eyes before washing your hands (Malakar, 2020).

Mainly, COVID-19 has been created a year of uncertainty for the world economy. Following discussion of each sector has proved the statement:

Economic turmoil related to the 2019–20 coronavirus pandemic has wide-ranging and severe impacts upon financial markets, including stock, bond, and commodity (including petroleum and gold) markets. Major events included a described Russia–Saudi Arabia oil price war after failing to reach an OPEC+ agreement that resulted in a collapse of crude oil prices and a stock market crash in March 2020. The effects upon markets are a part of the coronavirus recession and among the various socio-economic impacts of the pandemic.

Financial risk and country risk: As coronavirus put Europe and therefore the us in virtual lockdown, financial economists, credit rating and country risk experts have scrambled to rearrange their assessments in light of the unprecedented geo-economic challenges posed by the crisis. M. Nicolas Firzli, a director of the planet Pensions Council (WPC) and planning board member at the planet Bank Global Infrastructure Facility, refers thereto as "the Greater Financial Crisis" and says it's bringing to the surface many pent-up financial and geopolitical dysfunctions:

“So far, the sole European countries forced to place in situ short-selling bans are Italy, Spain and France: three of the four largest economies within the enfeebled European Union .” He believes the financial vulnerability of Madrid, Milan and Paris is thanks to an often-overlooked geo-economic reality, which could come to the fore within the coming days. “By OECD standards, Spain, Italy and France have very weak pension assets bases. Their combined pension wealth is quite 15 times smaller than that of jurisdictions like the united kingdom or Australia. In times of acute crisis, like today, they lack cash-rich domestic buyers of pis aller for the bonds and equities traded on their financial markets. Their national economies will suffer as a result, and their political sovereignty itself could also be severely eroded.” At the international and national levels, however—as Helmut Ettl, head of the Austrian financial market authority, said—there is not any reliable empirical data to measure the continued effects of the COVID-19 disease on the economy and therefore the environment, as this sort of crisis is unprecedented. Companies that were already financially weak before the crisis are now further destabilized. All that's known, Ettl said, is that the crisis are going to be profound.

Shopping and spread of the coronavirus: Individuals who have continued to buy during the 2019-20 coronavirus pandemic are at an increased risk of contracting COVID-19. Amidst the pandemic, grocery stores and pharmacies still remain open and attract crowds of shoppers, thus creating the potential to further spread contagion. Shoppers using folding money to form purchases may further be disseminating the coronavirus, consistent with the planet Health Organisation. Bank branches that still remain open amidst the 2019-20 coronavirus pandemic pose a health risk because the gathering of individuals allows for the spread of contagion.

Corporations' response to the health risk: As required by regulation, large American banks are reviewing emergency plans to make sure that they will still operate if conditions worsen; contingency plans include having employees work from their homes, transferring staff to backup offices and even sending staff to offices in other cities. Goldman Sachs cancelled all "non-essential" travel and put in situ specific restrictions on visit China, South Korea , Italy, and Iran. Amazon, who confirmed an employee within the Seattle area tested positive for COVID-19, banned nonessential employee travel.

Large technology companies have also made changes. Twitter "suspend all non-critical business travel and events". Facebook cancelled its F8 developer conference. Some companies, including Microsoft, GitHub, and Square, Inc., have also encouraged employees to figure from home. Other major corporations have reduced their commercial activity or otherwise modified their business practices. Nike closed its Oregon headquarters and its European headquarters in Amsterdam "out of an abundance of caution".

Stock Market: On Monday, 24 February 2020, the Dow Jones Industrial Average and FTSE 100 dropped quite 3% because the coronavirus outbreak spread worsened substantially outside China over the weekend. This follows benchmark indices falling sharply in continental Europe after steep declines across Asia. The DAX, CAC 40 and IBEX 35 each fell by about 4% and therefore the FTSE MIB fell over 5%. There was an outsized fall within the price of oil and an outsized increase within the price of gold, to a 7-year high. On 27 February, thanks to mounting worries about the coronavirus outbreak, various U.S. stock exchange indices including the NASDAQ-100, the S&P 500 Index, and therefore the Dow Jones Industrial Average posted their sharpest falls since 2008, with the Dow falling 1,191 points, its largest one-day drop since the 2008 financial crisis. On 28 February 2020, stock markets worldwide reported their largest single-week declines since the 2008 financial crisis. Following a second week of turbulence, on 6 March, stock markets worldwide closed down (although the Dow Jones Industrial Average, NASDAQ Composite, and S&P 500 closed abreast of the week), while the yields on 10-year and 30-year U.S. Treasury securities fell to new record lows under 0.7% and 1.26% respectively. U.S. President Donald Trump signed into law an emergency appropriations and pandemic countermeasures bill including $8.3 billion in government spending. After OPEC and Russia did not agree on boring cuts on 5 March and Saudi Arabia and Russia both announced increases in boring on 7 March, oil prices fell by 25 percent.

Oil prices: The reduction within the demand for travel and therefore the lack of factory activity thanks to the outbreak significantly impacted demand for oil, causing its price to fall. In mid-February, the International Energy

Agency forecasted that oil demand growth in 2020 would be the littlest since 2011. Chinese demand slump resulted during a meeting of the Organization of Petroleum Exporting Countries (OPEC) to debate a possible cut in production to balance the loss in demand. The cartel initially made a tentative agreement to chop boring by 1.5 million barrels per day (bpd) following a gathering in Vienna on 5 March 2020, which might bring the assembly levels to rock bottom it's been since the Iraq War. Meanwhile, analytics firm IHS Markit predicted a fall global demand for crude to fall by 3.8 million bpd within the half-moon of 2020, largely thanks to the halt to Chinese economic activity thanks to the virus; it also predicted the primary annual reduction in demand for crude since the financial crisis of 2007–08.

Bond market: Prior to the coronavirus pandemic, a huge amount of borrowing by firms with ratings just above "junk," including the expansion of leveraged loans, which are made to companies with significant amount of debt, created a vulnerability within the economic system . The collapse of this corporate debt bubble would potentially endanger the solvency of firms, potentially worsening subsequent recession. In January, new U.S. corporate debt fell 10% from the previous year, potentially indicating more caution from investors. because the economic impact of the coronavirus began to be felt, numerous financial news sources warned of the potential cascade of impacts upon the outstanding $10 trillion in corporate debt. Between mid-February and early March, investors increased the premium, or additional yield, to carry junk bonds by fourfold the premium demanded of upper credit lenders, indicating increased wariness.

So many measurement sources is says that, Covid-19 impact: Global economy to fall into recession in 2020, like, RBI says that, “RBI said growth would depend how fast the outbreak is contained and economic activity returns to normalcy. RBI expects domestic inflation for fiscal 2021 to move in the range of 3.6-3.8%.” therefore, COVID-19 has been created a year of uncertainty for the world economy.

Bibliography:

1) American society for Microbiology, (2020). https://asm.org/Press-Releases/2020/COVID-19-Resources

2) Bloomberg Opinion (2020). Coronavirus Exposes India's Official Callousness0 (https://www.bloombergquint.com/global-economics/india-s-government-should-have-expected-coronavirus-migration).
3) Borter, Gabriella; Gorman, Steve (March 6, 2020). "Coronavirus found on cruise ship as more U.S. states report cases". Reuters. Thomson Reuters. Archived from the original on March 10, 2020.
4) Business Today (2020). Why suspension of MPLADS is problematic in corona times? (https://www.businesstoday.in/current/economy-politics/coronavirus-in-india-why-suspension-of-mplads-is-problematic/story/400771.html).
5) CDC (https://www.cdc.gov/coronavirus/2019-ncov/specific-groups/high-risk-complications.html)
6) Census of India (https://censusindia.gov.in/)
7) Coronavirus: Microsoft, Square, Twitter encourage employees to work from home". San Francisco Chronicle. 5 March 2020.
8) Deutsch, Anthony (2020-03-02). "UPDATE 2-Nike temporarily closes European HQ in the Netherlands due to coronavirus case". CNBC.
9) Dev, S.M., (2020). Addressing COVID-19 impacts on agriculture, food security, and livelihoods in India. International Food Policy Research Institute, IFPRI blog pages. (https://www.ifpri.org/blog/addressing-covid-19-impacts-agriculture-food-security-and-livelihoods-india).
10) Dow Industrials Drop 1,000 Points as Coronavirus Cases Mount Outside Asia". The Wall Street Journal. Archived from the original on 24 February 2020.
11) Franck, Thomas (March 6, 2020). "10-year Treasury yield falls to new all-time low under 0.7% as flight to bonds continues". CNBC. Archived from the original on March 6, 2020.
12) Global stock markets plunge on coronavirus fears". Archivedfrom the original on 25 February 2020.
13) Govt. of India (COVID-19 database), (26th April, 2020) https://www.covid19india.org/

14) Huang, Eustance (March 6, 2020). "Major Asian markets fall more than 2% to cap week of market gyrations". CNBC. Archivedfrom the original on March 6, 2020.

15) Imbert, Fred; Huang, Eustance (February 27, 2020). "Dow falls 350 points Friday to cap the worst week for Wall Street since the financial crisis". CNBC. Archived from the original on 28 February 2020.

16) Johnson, Keith (5 March 2020). "OPEC Tries to Forestall a Coronavirus Oil Collapse". Foreign Policy. Archived from the original on 7 March 2020. Lynch, David J. (10 March 2020). "Fears of corporate debt bomb grow as coronavirus outbreak worsens". Washington Post. Archived from the original on 11 March 2020.

17) Kelly, Stephanie (March 8, 2020). "Oil plunges 25%, hit by erupting Saudi-Russia oil price war". Reuters. Thomson Reuters. Archived from the original on March 9, 2020.

18) Kollewe, Julia (4 February 2020). "Opec discusses coronavirus as Chinese oil demand slumps – as it happened". The Guardian. Archived from the original on 6 February 2020.

19) Li, Yun (March 6, 2020). "Dow falls 250 points, ekes out a small weekly gain after a wild run". CNBC. Archived from the original on March 6, 2020.

20) Malakar. K.D., (2020). Coronavirus (COVID-19): A Pandemic Situation of in the World. Lambert Academic Publishing, Germany. ISBN: 978-620-2-51827-7.

21) McCaffrey, Orla (19 March 2020). "As Coronavirus Spreads, Banks Face a Tough Call on Branch Closures". Wall Street Journal.

22) Oil prices fall as coronavirus spreads outside China". AP. 29 February 2020. Archived from the original on 8 March 2020. Coronavirus set to knock oil demand growth to slowest since 2011". Financial Times. 13 February 2020. Archived from the original on 14 February 2020.

23) Politics and Nations (2020). More than 21,000 camps set up for over 6,60,000 migrants: State governments. (https://economictimes.indiatimes.com/news/politics-and-nation/more-than-21000-camps-set-up-for-over-660000-migrants-state-governments/articleshow/74920798.cms)

24) Ponczek, Sarah; Hajric, Vildana (9 March 2020). "S&P 500 Plunges 7%, Triggering Market-Wide Stock Trading Halt". Bloomberg. Archived from the original on 9 March 2020.

25) Ramano, Benjamin (3 March 2020). "Amazon employee in Seattle has tested positive for illness caused by coronavirus". The Seattle Times.

26) Sangster, Kalila (3 March 2020). "Banknotes may be spreading coronavirus, World Health Organisation warns". Yahoo!

27) Schrotenboer, Brent (27 March 2020). "Are grocery stores and pharmacies vectors for the coronavirus?". USA TODAY.

28) Smith, Elliot (February 28, 2020). "European stocks fall 12% on the week as coronavirus grips markets". CNBC. Archived from the original on 28 February 2020.

29) Smith, Elliot (February 28, 2020). "Global stocks head for worst week since the financial crisis amid fears of a possible pandemic". CNBC. Archived from the original on 28 February 2020.

30) Smith, Elliot; Amaro, Silvia (March 6, 2020). "European stocks close 3.6% lower amid coronavirus volatility; oil sector down 5.5%". CNBC. Archived from the original on March 6, 2020.

31) Stevens, Pippa; Meredith, Sam (March 6, 2020). "Oil plunges 10% for worst day in more than 5 years after OPEC+ fails to agree on a massive production cut". CNBC. Archived from the original on March 8, 2020.

32) Tappe, Anneken (27 February 2020). "Dow falls 1,191 points -- the most in history". CNN Business. CNN. Archived from the original on 28 February 2020.

33) The Atlantic (2020). The Callousness of India's COVID-19 Response. (https://www.theatlantic.com/international/archive/2020/03/india-coronavirus-covid19-narendra-modi/608896/).

34) The Dow Is Down 700 Points as the Coronavirus Strikes in Italy". Barrons. Barrons. Archived from the original on 24 February 2020.

35) The INDIAN Express (2020). Union Government should rethink Central Vista project.

(https://www.newindianexpress.com/opinions/2020/apr/25/union-government-should-rethink-central-vista-project-2134979.html).

36) The Print (2020). Rotis, mobile recharges, carrom boards — how Kerala fixed its migrant worker anger. (https://theprint.in/india/rotis-mobile-recharges-carrom-boards-how-kerala-fixed-its-migrant-worker-anger/403937/)

37) The Times of India (2020). Delhi lockdown: Over 500 hunger relief centres set up for 4 lakh people. (https://timesofindia.indiatimes.com/city/delhi/govt-to-give-2-meals-a-day-to-4-lakh-people-says-kejriwal/articleshow/74855058.cms).

38) The Tribune (2020). Over 75 lakh being fed at food camps: MHA. (https://www.tribuneindia.com/news/nation/over-75-lakh-being-fed-at-food-camps-mha-66223).

39) The week (2020). COVID-19: Opposition fumes over Modi govt decision to suspend MPLADS funds. (https://www.theweek.in/news/india/2020/04/06/opposition-parties-hit-out-at-govt-over-mplads-decision-say-will-hit-development-works.html).

40) Twitter, Amazon.com, Nike and more take emergency coronavirus precautions". Yahoo! Finance.

41) WHO (https://www.who.int/emergencies/diseases/novel-coronavirus-2019)

42) Wiseman, Paul; Condon, Bernard; Bussewitz, Cathy (11 March 2020). "Corporate debt loads a rising risk as virus hits economy". AP News. Archived from the original on 12 March 2020. Oliver, Matt (10 March 2020). "The £15trillion corporate debt time bomb: Global economy at risk from the bond market as virus spreads". MSN.

The End

Chapter – 11

Case Study-iii

Management of COVID-19's Health Crisis: A study of Maharashtra, India

Kousik Das Malakar[1]

Abstract:

Aims and Objectives: The present study focuses on determining the status of COVID-19's Health Crisis and management in Maharashtra, India. This study seeks to examine that, (a) evaluate the changing situation of COVID-19 in Maharashtra, and (b) Define the various types of health-related crises and management programs in Maharashtra's health sectors.

Data and Methods: Data set compiled from various secondary sources like Indian state-wise COVID-19 data from Online COVID-19 (2020), Govt. of India; Public Health Department, Maharashtra (2020) and Arogya, Maharashtra (2020). And using the MS Excel 2014 & ArcGIS 10.2.2 applications.

Results: India as well as Maharashtra's COVID-19 affected cases has been increased day by day. And it's created health crisis and related socio-economic crisis also. The present study try to focus on the various management funda (programs) in Maharashtra's health sectors.

Keywords: COVID-19; Health crisis; Health management; Socio-economic crisis; Maharashtra

----

[1]Centre for the Study of Regional Development (CSRD), School of Social Sciences (SSS), Jawaharlal Nehru University (JNU), New Delhi, India. E-mail: kdmalakargegr92@gmail.com

1. Introduction:

COVID-19 is a pneumonia-like disease with symptoms including fever, dry cough, Tightness of Chest, Head Ache, shortness of breath and Kidney Failure that was first identified in Wuhan, China (American society for Microbiology, 2020 and WHO, 2020). Mainly, older adults and People who have serious chronic medical conditions like, Heart disease Diabetes Lung disease suppressed immune systems, whose are more affected than the others (CDS, 2020). COVID-19 is transmitted human to human by the medium of air by coughing and sneezing, Close personal contact, such as touching or shaking hand, touching an object or surface with the virus on it, then touching your mouth, nose, or eyes before washing your hands (Malakar, 2020).

In India, the first case of COVID-19 was reported on January 30, 2020. As of May 12, confirmed cases of COVID-19 stand at 71,276. While 45,926 patients have active cases, 23,028 have recovered, and 2,310 have died (Govt. of India, 12nd May, 2020). India has taken early action to limit the spread of COVID-19, ordering a 21-day nationwide lockdown for its population of 1.3 billion people starting March 25. The novel coronavirus has spread widely in India relatively recently compared to other countries, and the number of reported infections is low so far, with 5,274 cases and 149 deaths as of April 8. However, as COVID-19 cases are increasing fast, there's great concern about the disease's potential spread and impact. India has to be ready for a possible surge. Testing should be expanded significantly. The government views the pattern of the spread of COVID-19 as almost like the 2009 H1N1 influenza pandemic, meaning the spread is unlikely to be uniform. After the 21-day period expires, it is planning to maintain the full lockdown in "hotspot" areas and relax it in other places (Dev, 2020). And currently, lockdown conditions is growing up.

These measures may help in limiting the health crisis, but as in other countries, the complete shutdown of all economic activities except essential services will create an economic crisis and misery for the poor, with massive job losses and rising food insecurity.

The economic shock will likely be much more severe for India, for two reasons. First, pre-COVID-19, the economy was already slowing down, compounding existing problems of unemployment, low incomes, rural distress, malnutrition, and widespread inequality. Second, India's large informal sector is particularly vulnerable. Out of the national total of 465 million workers, around 92.8% were informal workers in 2017-18 (Census of India, 2018). Lacking regular salaries or incomes, these agriculture, migrant, and other informal workers would be hardest-hit during the lockdown period. Mainly, its fall the effects on various socio-economic activities (The Ministry of Labour & Employment, Govt. of India, 2020) as well as climatic phenomenon (Central Pollution Control Board, Govt. of India, 2020) also.

In this paper focuses the main problems of COVID-19 in Maharashtra according to the date of 12nd May, 2020 (perspectives of health crisis). Here, focus on the impacts on health crisis and that type of crisis fall the effect on another importance sectors in Maharashtra, like, Education, transport, tourism, marginal workers and economic effects, etc.

2. Objectives:

Therefore, the specific objectives of this study are;

1) To evaluate the changing situation of COVID-19 in Maharashtra.
2) Define the various types of health-related crises and management programs in Maharashtra's health sectors.

3. Study area:

The study area is Maharashtra. Situated in the western part of the country, the Deccan, Maharashtra happens to be the third-largest state in the nation and stands second in population among all Indian states. Surrounded by the Arabian Sea to its west, Gujarat to its northwest, Madhya Pradesh to its north, Karnataka to its south and Chhattisgarh and Telangana to its East, this state of India is 3,07,713 sq km in area. Mumbai is the capital of Maharashtra as well as the financial capital of the entire country. Nagpur is known as the auxiliary capital of the state. Also known as the wealthiest state, Maharashtra contributes around 15% of the industrial output of the country and around 14% of its gross domestic product.

Maharashtra experiences a tropical monsoon type of climate. The annual rainfall of the state can vary from 400 - 6000 mm and occurs for 3 - 4 months in a year. The average annual temperature of Maharashtra remains 25 - 27 degrees centigrade in most regions.

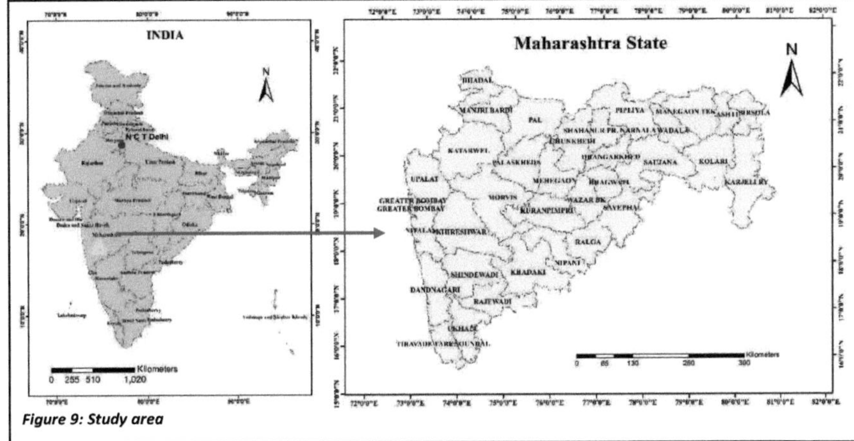

Figure 9: Study area

4. Database and Methodology:

For the present study data set compiled from various secondary sources like, [2]Indian state-wise COVID-19 data from Online COVID-19 (2020), Govt. of India; [3]Public Health Department, Maharashtra (2020) and [4]Arogya, Maharashtra (2020).

---

[2]https://www.covid19india.org/

[3]https://phdmah.maps.arcgis.com/

[4]https://arogya.maharashtra.gov.in/

Also calculate the data by using MS Excel format 2014 and ArcGIS 10.2.2 and prepared the mapping used the various types of techniques of ArcGIS application 10.2.2 and figurate the bar diagram using the MS Excel format 2014.

5. About COVID-19:

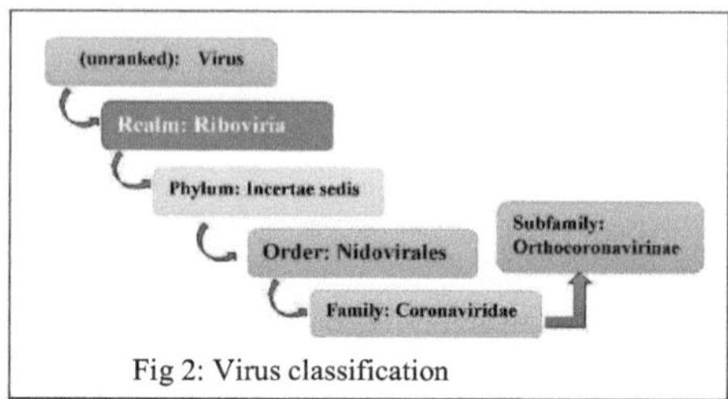

Fig 2: Virus classification

Coronaviruses constitute the subfamily Orthocoronavirinae, within the family Coronaviridae, order Nidovirales, and realm Riboviria. They have enveloped viruses with a positive-sense single-stranded RNA genome and a nucleocapsid of helical symmetry. The genome size of coronaviruses ranges from approximately 27 to 34 kilobases, the most important among known RNA viruses. The name coronavirus springs from the Latin corona, meaning "crown" or "halo", which refers to the characteristic appearance like a crown or a solar corona around the virions (virus particles) when viewed under two-dimensional transmission microscopy, thanks to the surface being covered in club-shaped protein spikes.

Various Names:

Firstly, COVID-19 also known as, Coronavirus disease 2019, Novel coronavirus pneumonia, 2019-nCoV acute respiratory disease, and Wuhan coronavirus also.

Symptoms:

COVID-19 symptoms range from mild to severe. It takes 2-14 days after exposure for symptoms to develop. Symptoms may include:

• Fever • Difficulty in Breathing • Coughing • Tightness of Chest • Running Nose • Head Ache • Feeling of being Unwell • Pneumonia • Kidney Failure.

Risk Persons:

According to the CDC, children don't seem to be at higher risk for getting COVID-19. However, some people are, including Older adults and People who have serious chronic medical conditions like: Heart disease Diabetes Lung disease Suppressed immune systems.

Causes a Coronavirus Infection:

Humans rst get a coronavirus from contact with animals. Then, it can spread from human to human. Health ocials do not know what animal caused COVID-19.

The COVID-19 virus can be spread through contact with certain bodily uids, such as droplets in a cough. It might even be caused by touching something an infected person has touched then touching your hand to your mouth, nose, or eyes.

Mode of transmission:

Human Coronavirus (2019-nCoV) most ordinarily spread from an infected person to other through:

- The air by coughing and sneezing.
- Close personal contact, such as touching or shaking hand.
- Touching an object or surface with the virus on it, then touching your mouth, nose, or eyes before washing your hands.

6. Result and Discussion:

6.1 Present Situation of India:

As per the date of 12nd May 2020, India's affected due to COVID-19 are more than 26, 000. Most of vulnerable state is Maharashtra (More than 23,000), than Gujarat (More than 8,500), Delhi (More than 7,500), Rajasthan (More than 4,000), Madhya Pradesh (More than 3,700), Tamil Nadu (More

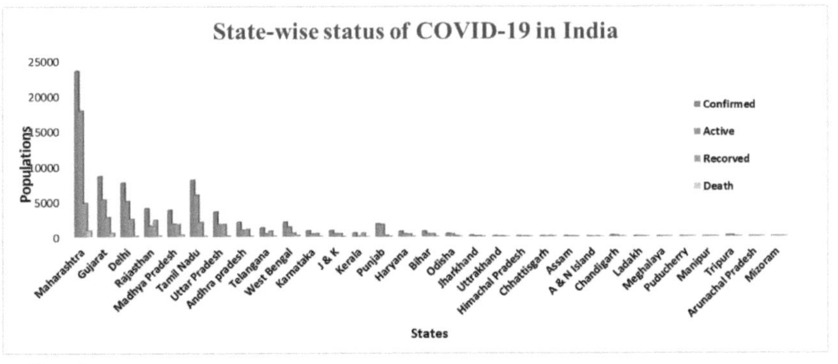

Figure 3

than 8,000) and others. Following figure 3 shows the status of state-wise COVID-19's situations. (Source: COVID-19, Govt. of India (https://www.covid19india.org/)).

6.2 Present Situation of Maharashtra:

As per the date of 12nd May 2020, India's affected due to COVID-19 are more than 26, 000. Most of the vulnerable districts is Mumbai (More than 14,500), than Pune (More than 2,700), Thane (More than 2,500), Nashik (More than 696), Aurangabad (More than 580), and others. Following figure

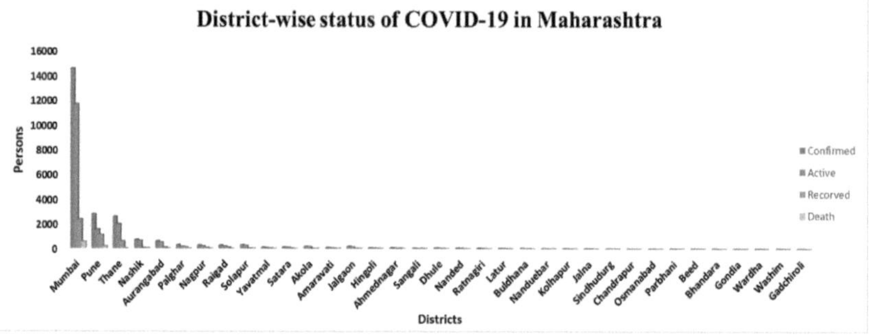

Figure 4

4 shows the status of district-wise COVID-19's situations in Maharashtra. (Source: COVID-19, Govt. of India (https://www.covid19india.org/)).

Following charts (Fig 5) shows the, at a glances configuration of COVID-19's conditions of Maharashtra (Source: Ministry of Health and Family Welfare, Govt. of India).

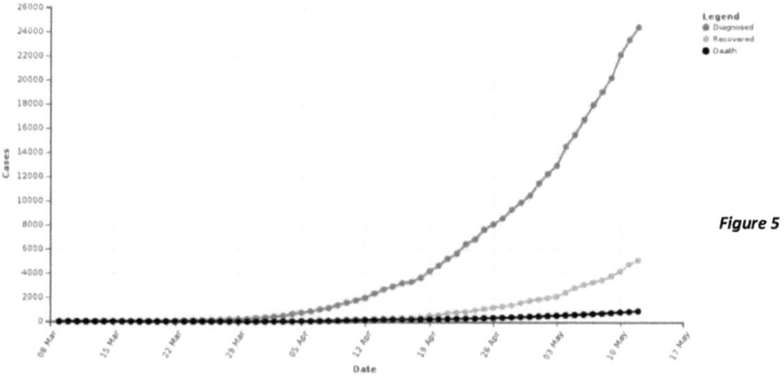

Figure 5

Following charts (Fig 6) shows the Daily new cases in Maharashtra.

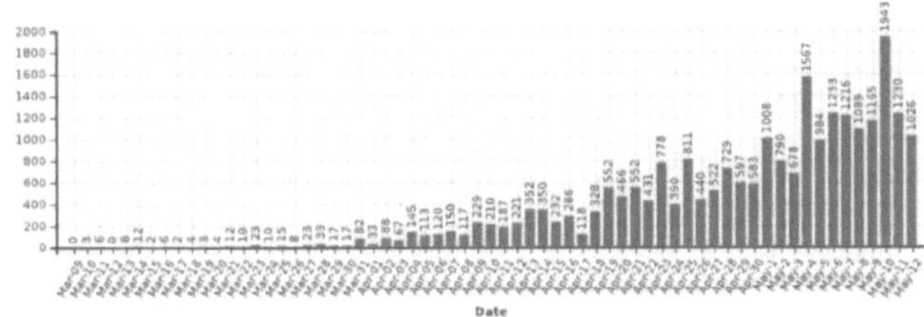

Figure 6

Following charts (Fig 7) shows the Daily new recoveries in Maharashtra.

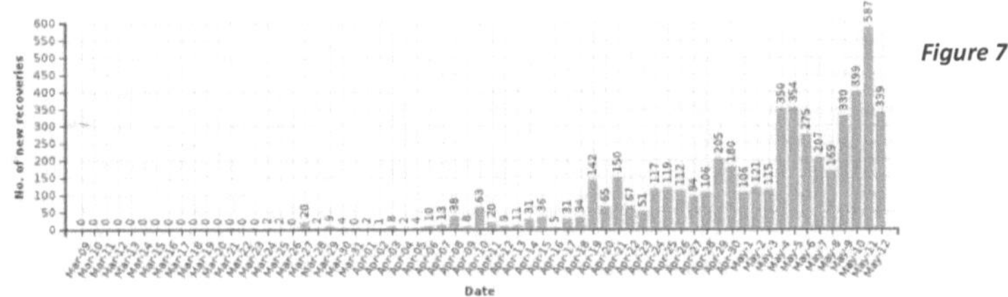

Figure 7

Following charts (Fig 8) shows the Daily new deaths in Maharashtra.

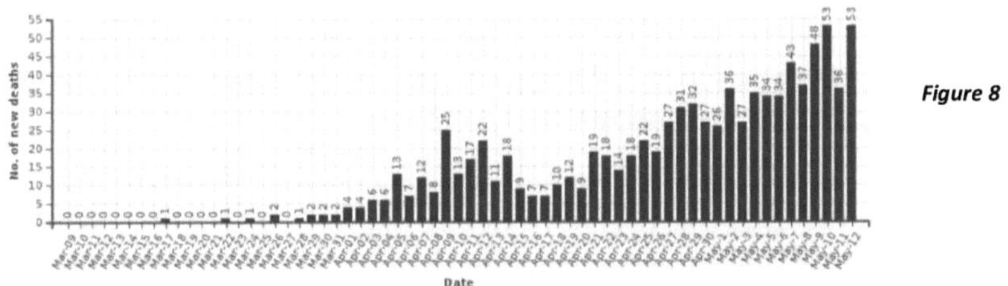

Figure 8

Following charts (Fig 9) shows the Semilog plot of the spread of SARS-CoV-2 and of COVID-19 recoveries & deaths in Maharashtra

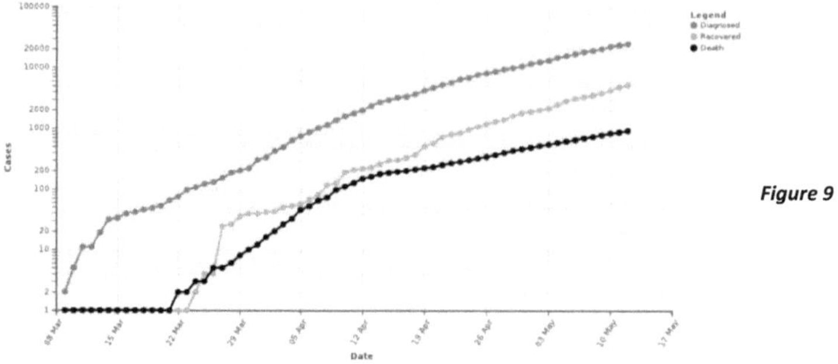

Figure 9

6.2 Impacts and managements:

6.2.1 Transport

Over 20,000 bus services of Maharashtra State Road Transport Corporation were cancelled since 11 March, which caused the organization losses of ₹3 crores by 17 March. Indian Railways cancelled 23 trains from the state in order to prevent the spread of the virus to other states. The government of Gujarat started screening people entering from Mumbai and other borders with Maharashtra by mid of March. The state border was sealed completely during the first lockdown. Since Mumbai started facing pandemic before Gujarat these precautionary measures were being taken as soon as the first lockdown began. The government of Telangana implemented the screening of people entering the state of Telangana from Maharashtra at four entry points. The government of Madhya Pradesh suspended all bus services between Indore and Maharashtra until 31 March.

On 22 March, the Indian Railways announced that the Mumbai Suburban Railway will be closed between 22 and 31 March. Mumbai Monorail and Mumbai Metro services were also cancelled until the end of the month. Thackeray announced that state-run and private bus services will be suspended for the general public until further notice.

6.2.2 Tourism

Tourist spots in the state like Ajanta and Ellora Caves in Aurangabad district, Elephanta Island in Raigad district, and Gateway of India in Mumbai reportedly witnessed a sharp decline in the number of visitors. Hotel, cab, and private bus businesses in the state also reported a high number of booking cancellations in March.

Amidst a surge in confirmed cases across the state, the health officials declared that several tourist and non-secular sites are going to be closed down as a precautionary measure. These sites included Siddhivinayak Temple in Mumbai, Tulja Bhavani Temple in Osmanabad district, Ajanta and Ellora Caves in Aurangabad district, Dagadusheth Halwai Ganapati Temple in Pune, Mumba Devi Temple in Mumbai and Saibaba Temple in Shirdi. Entry restrictions were also enforced at Mantralaya, Mumbai. Mumbai Police placed a ban on all group tours within the city until 31 March.

6.2.3 Education

Maharashtra government cancelled all the exams from standard 1–8. to make it easier to stop the Coronavirus outbreak through schools in students.The students of standard 1st to 8th standard were directly promoted to their next standard.

6.2.4 Migrant workers

After the enforcement of lockdown, scores of migrant labourers in Maharashtra who were left without work began to leave for their respective states. Thousands of workers reportedly started going to Gujarat and Rajasthan by foot as transport facilities were suspended, but were stranded at the state border. Thackeray appealed to the migrant workers to not travel back to their states and announced a ₹ 45 crore package to provide food and accommodation for migrant labourers. On 30 March, the state government set up 262 relief camps across the state, providing food and shelter to 70,339 migrant workers and homeless people. On 8 April, it was reported that 4,653 camps were functioning across the state, which housed more than 550,000 people. These camps were started by district administrations, labour department, irrigation department, and cooperative sugar mills. Several employers such as Larsen & Toubro, Godrej Group and Shapoorji Pallonji announced that they would provide food, accommodation, and wages to their migrant labourers.

On 14 April, thousands of migrant workers protested against the lockdown near Bandra railway station in Mumbai and demanded to be sent back to their states. Police made use of baton charges to disperse the unruly crowd and control the situation.

6.2.5 Economic effects

The virus outbreak had a significant impact on the automobile sector in the state, according to a CNBC TV18 article on 21 March. Bajaj Auto closed down its manufacturing facilities in Akurdi and Chakan until 30 March, while Tata Motors scaled down operations in its Pune plant. Eicher Motors and Ashok Leyland shut down components plant in Thane and Bhandara respectively. Mercedes-Benz suspended operations at its Chakan facility until 31 March. Fiat, Force Motors and JCB also announced the suspension of operations at units in Ranjangaon, Akurdi, and Chakan respectively, until 31 March. Mahindra & Mahindra also declared that it would suspend manufacturing at its Nashik plant and production at its Chakan and Kandivali units, starting from 23 March.

According to a 17 March article in the Indian Express, the economy of Mumbai was projected to suffer losses worth at least ₹ 16,000 crores per month in the service sector, as a result of the outbreak. In addition, it predicted that the city's tourism industry would lose ₹ 2,200 crores per month from international tourists.

The entertainment industry in the state also suffered financially, with several Bollywood films postponing release dates or halting production, causing an "acute financial crunch" for many members of the Federation of Western India Cine Employees. Bollywood films were estimated to lose box office revenue of over ₹ 1,300 crore due to the lockdown.

On 30 March, the state government slashed electricity tariff by an average of 8 percent for a period of five years in order to minimize the economic impact of the outbreak on industries. Deputy Chief Minister and Finance Minister, Ajit Pawar, announced that the March salaries of all elected representatives (MLAs, MLCs and Chief Minister) would be cut by 60 percent, while some of the government employees will also see 25 to 50 percent salary cuts due to the economic crisis arising from outbreak and lockdown.

7. Conclusion:

Finally, COVID-19 created a Pandemic situation in the world and its fall effects on Indian socio-economic sectors. And it is being created a negative growth rate of various fields. Presently, fight only COVID-19 and maintain physical distance, not a social distance (it's mandatory to each other contact remotely), and follow the government precautions. And take the actions for development and supply the proper health instruments, provide the educational infrastructure for future generation in this lockdown period, provided the packages of daily and marginal workers etc.

Acknowledgment

The author thank anonymous reviewer for constructive comments. And thanks to Dr. D.N.Das, Dr. Bhaswati Das, Dr. Milap Punia, Dr. Sanjeev Sharma (CSRD, JNU) and Dr. Nieuwaal (Director Delta Alliance, The Netherlands) for the theoretical and applied helps of this paper.

8. References:

1) American society for Microbiology, (2020). https://asm.org/Press-Releases/2020/COVID-19-Resources
2) Bollywood braces for huge losses amid coronavirus lockdown". Euronews. Archived from the original on 30 March 2020.
3) CDC (https://www.cdc.gov/coronavirus/2019-ncov/specific-groups/high-risk-complications.html)
4) Census of India (https://censusindia.gov.in/)
5) Central Pollution Control Board, Govt. of India (https://cpcb.nic.in/).
6) Choudhary, Abhishek (2020). "Maharashtra schools face Covid-19 test: No exams from std 1-8". The Times of India.
7) Containing Coronavirus: Maruti Suzuki, M&M, Hero MotoCorp Suspend Production". Bloomberg Quint. Archived from the original on 23 March 2020.
8) Coronavirus Fallout: From Maharashtra, an Exodus of Migrant Workers With No Work". The Wire.

9) Coronavirus impact: Automobile companies begin plant shutdowns in Maharashtra". CNBC TV18. Archived from the original on 22 March 2020.

10) Coronavirus lockdown: 3,000 migrant workers going on foot stuck at Maharashtra-Gujarat border". Deccan Herald.

11) Coronavirus Outbreak: Bus services between Indore and Maharashtra to be suspended". India TV. Archived from the original on 19 March 2020. Retrieved 18 March 2020.

12) Coronavirus pandemic | Car, bike factories in Pune shut till 31 March; Mercedes, Fiat, JCB join the league. Moneycontrol. Archived from the original on 22 March 2020.

13) Coronavirus pandemic: Maharashtra govt announces 8 per cent cut in electricity tariff for 5 years". The Times of India. Archived from the original on 31 March 2020.

14) Coronavirus scare hits hotel, transport business in Pune. The Times of India. Archived from the original on 12 March 2020.

15) COVID-19 impact: Mumbai economy staring at loss of around Rs 16,000 crore". The Indian Express. Archived from the original on 27 March 2020.

16) Covid-19 impact: Mumbai Police ban group tours in city till March 31". The Economic Times. Archived from the original on 31 March 2020.

17) COVID-19 pandemic in Maharashtra. 8 May 2020. Retrieved 8 May 2020.

18) Dev, S.M., (2020). Addressing COVID-19 impacts on agriculture, food security, and livelihoods in India. International Food Policy Research Institute, IFPRI blog pages. (https://www.ifpri.org/blog/addressing-covid-19-impacts-agriculture-food-security-and-livelihoods-india).

19) Govt. of India (COVID-19 database), (26th April, 2020) https://www.covid19india.org/

20) Local, Metro and long distance train services suspended till March 31". Mumbai Mirror. Archived from the original on 23 March 2020. Retrieved 23 March 2020.

21) Maharashtra announces salary cuts for State politicians, govt personnel". Business Line. Archived from the original on 4 April 2020.

22) Maharashtra earmarks Rs 45 crore for food, shelter to migrant workers". The New Indian Express.

23) Maharashtra government sets up relief camps for migrant workers". Pune Mirror. Retrieved 10 April 2020.

24) Maharashtra tourism is a casualty of coronavirus". The Times of India. Archived from the original on 14 March 2020. Retrieved 17 March 2020.

25) Maharashtra: Central railway cancels few long distance trains to curb spread of coronavirus". The Times of India. Archived from the original on 18 March 2020.

26) Malakar. K.D., (2020). Coronavirus (COVID-19): A Pandemic Situation of in the World. Lambert Academic Publishing, Germany. ISBN: 978-620-2-51827-7.

27) Restrictions on Maha border extended by T". The Times of India. Archived from the original on 18 March 2020.

28) Shirdi Saibaba Temple to shut from today amid coronavirus scare". Livemint. Archived from the original on 17 March 2020.

29) The Ministry of Labour & Employment, Govt. of India (https://labour.gov.in/)

30) Thousands of migrant labourers assemble outside Mumbai station hoping to reach homes". Livemint. Retrieved 16 April 2020.

31) Three Maharashtra labs testing coronavirus samples from other states too. The Economic Times. Retrieved 6 April 2020.

32) Uddhav Thackeray appeals to migrant workers in Maharashtra to stay put". Business Line.

33) WHO (https://www.who.int/emergencies/diseases/novel-coronavirus-2019)

34) Why isn't Mumbai seeing an exodus of migrant workers?. Livemint. Retrieved 10 April 2020.

The End

Chapter - 12

Bibliography

1) A Dictionary of Epidemiology (IEA)
2) A History of Spain. Archived from the original on 27 March 2017. Retrieved 16 November 2014.
3) A History of the Human Plague in Iran Archived 2011-07-19 at the Wayback Machine, Mohammad Azizi, Farzaneh Azizi
4) A Prospective/Retrospective,Randomized Controlled Clinical Study of Antiviral Therapy in the 2019-nCoV Pneumonia - Full Text View - ClinicalTrials.gov . clinicaltrials.gov. Retrieved 20 March2020.
5) A real-world study for lopinavir/ritonavir (LPV/r) and emtritabine (FTC) / Tenofovir alafenamide Fumarate tablets (TAF) regimen in the treatment of novel coronavirus pneumonia (COVID-19) . chictr.org.cn (in Chinese). Retrieved 20 March 2020.
6) Abbott, Alison (17 June 2009). Italy cancels G8 research meeting . Nature. doi:10.1038/news.2009.579. ISSN 0028-0836.
7) Abdul Hadi, Eddino (30 January 2020). Wuhan virus: Two K-pop concerts in Singapore postponed . The Straits Times. Retrieved 24 February 2020.
8) Aboriginal Health History. Archived from the original on 27 September 2011. Retrieved 16 November 2014.
9) Abueish, Tamara (24 February 2020). First case of coronavirus detected in Bahrain . Al Arabiya. Retrieved 24 February 2020.
10) Actualización nº53. Enfermedad por el coronavirus (COVID-19)(PDF) (Report) (in Spanish). Ministerio de Sanidad, Consumo y Bienestar Social. 23 March 2020. Retrieved 24 March 2020.
11) Acuna-Soto, R.; Romero, L. C.; Maguire, J. H. (2000). Large epidemics of hemorrhagic fevers in Mexico 1545–1815 . The American Journal of Tropical Medicine and Hygiene. 62 (6): 733–739. doi:10.4269/ajtmh.2000.62.733.
12) Acuna-Soto, Rodolfo; Stahle, D. W.; Cleaveland, M. K.; Therrell, M. D. (2002). Megadrought and Megadeath in 16th Century Mexico . Emerging Infectious Diseases. 8 (4): 360–362. doi:10.3201/eid0804.010175.

13) Advice for public . World Health Organization (WHO). Archived from the original on 26 January 2020. Retrieved 25 February 2020.

14) Advice for public . World Health Organization (WHO). Archived from the original on 26 January 2020. Retrieved 15 February 2020.

15) AFP (19 March 2020). Updated: WHO Now Doesn't Recommend Avoiding Ibuprofen For COVID-19 Symptoms . ScienceAlert. Retrieved 19 March 2020.

16) African trypanosomiasis Archived 2016-12-04 at the Wayback Machine, World Health Organization

17) After chaos, Dr Li Wenliang, coronavirus whistle-blower, confirmed dead at 34 . South China Morning Post. 7 February 2020. Archived from the original on 7 February 2020. Retrieved 8 February 2020.

18) Ahrens, Kevin; Ingvorsen, Emil Søndergård (29 February 2020). Tredje dansker smitttet med coronavirus [Third Dane infected with coronavirus]. DR (in Danish). Retrieved 29 February 2020.

19) AIFA e Gilead annunciano che l'Italia è tra i Paesi che testeranno l'antivirale remdesivir per il trattamento del COVID-19 . aifa.gov.it(in Italian). Retrieved 19 March 2020.

20) AIFA precisa, uso favipiravir per COVID-19 non autorizzato in Europa e USA, scarse evidenze scientifiche sull'efficacia . aifa.gov.it (in Italian). Retrieved 23 March 2020.

21) Air Canada, other airlines suspend flights to China over coronavirus fears . Global News. 29 January 2020. Retrieved 29 January 2020.

22) Akhtar, Tanzeel (6 February 2020). WHO Says There's No Effective Coronavirus Treatment Yet . Archived from the original on 6 February 2020. Retrieved 8 February 2020.

23) Ali, Imtiaz. Pakistan confirms first two cases of coronavirus, govt says 'no need to panic' . Dawn. Retrieved 26 February 2020.

24) Almeida JD, Berry DM, Cunningham CH, Hamre D, Hofstad MS, Mallucci L, McIntosh K, Tyrrell DA (1968). Coronaviruses . Nature. 220 (5168): 650. doi:10.1038/220650b0. PMC 7086490. [T]here is also a characteristic fringe of projections 200 A long,

which are rounded or petal shaped ... This appearance, recalling the solar corona, is shared by mouse hepatitis virus and several viruses recently recovered from man, namely strain B814, 229E and several others.

25) Almost 60 reported cases of coronavirus in Germany: Health ministry . Al Arabiya. 28 February 2020. Retrieved 28 February2020.

26) Alwan A, Mahjour J, Memish ZA (2013). Novel coronavirus infection: time to stay ahead of the curve . Eastern Mediterranean Health Journal. 19 Suppl 1: S3-4. doi:10.26719/2013.19.supp1.S3.

27) Ambrosevideo.com. Archived from the original on 2016-04-17. Retrieved 2011-03-27.

28) American Indian Epidemics. Archived from the original on February 14, 2015.

29) American plague. New Scientist. December 19, 2000. Archived from the original on October 17, 2018. Retrieved October 17, 2018.

30) Amid Coronavirus Outbreak, Air India To Suspend Flights To Hong Kong From February 8: Report . NDTV.com. Archivedfrom the original on 6 February 2020. Retrieved 6 February 2020.

31) Amid Ongoing COVID-19 Pandemic, Governor Cuomo Accepts Recommendation of Army Corps of Engineers for Four Temporary Hospital Sites in New York . governor.ny.gov. 22 March 2020.

32) Andersen KG, Rambaut A, Lipkin WI, Holmes EC, Garry RF (17 March 2020). The proximal origin of SARS-CoV-2 . Nature Medicine: 1–3. doi:10.1038/s41591-020-0820-9. Retrieved 18 March 2020.

33) Andersen, P. I.; Ianevski, A.; Lysvand, H.; Vitkauskiene, A.; Oksenych, V.; Bjørås, M.; Telling, K.; Lutsar, I.; Dumpis, U.; Irie, Y.; Tenson, T.; Kantele, A.; Kainov, D. E. (2020). Discovery and development of safe-in-man broad-spectrum antiviral agents . International Journal of Infectious Diseases : Ijid : Official Publication of the International Society for Infectious Diseases. 93: 268–276. doi:10.1016/j.ijid.2020.02.018.

34) Anderson RM, Heesterbeek H, Klinkenberg D, Hollingsworth TD (March 2020). How will country-based mitigation measures influence the course of the COVID-19 epidemic? . Lancet.

395(10228): 931–934. doi:10.1016/S0140-6736(20)30567-5. A key issue for epidemiologists is helping policy makers decide the main objectives of mitigation—e.g. minimising morbidity and associated mortality, avoiding an epidemic peak that overwhelms health-care services, keeping the effects on the economy within manageable levels, and flattening the epidemic curve to wait for vaccine development and manufacture on scale and antiviral drug therapies.

35) Anderson RM, Heesterbeek H, Klinkenberg D, Hollingsworth TD (March 2020). How will country-based mitigation measures influence the course of the COVID-19 epidemic?. Lancet. 395(10228): 931–934. doi:10.1016/S0140-6736(20)30567-5. Andramuño, Catalina (31 January 2020). Coronavirus: Ecuador cuenta con reactivos para detectar casos, anunció ministra de Salud (in Spanish). Retrieved 31 January 2020.

36) Andrew Davidson (1893). Hygiene & diseases of warm climates. Pentland. p. 337. Retrieved 31 March 2011.

37) Andrew Dunn (24 March 2020). The first study of a potential coronavirus vaccine will soon start recruiting healthy volunteers in Georgia at Emory University . Business Insider. Retrieved 24 March 2020.

38) Andrew Ekonomou. Byzantine Rome and the Greek Popes. Lexington Books, 2007

39) Andrew Osborn (20 March 2020). Russia starts testing coronavirus vaccine prototypes on animals . U.S. News & World Report. Thomson Reuters. Retrieved 21 March 2020.

40) Ang, Benson (26 January 2020). Wuhan virus: Andy Lau cancels all his 12 concerts in Hong Kong; Leon Lai postpones Macau shows . The Straits Times. Retrieved 24 February 2020.

41) Ang, Benson (28 January 2020). Wuhan virus: Miriam Yeung's Feb 8 Singapore concert postponed until further notice . The Straits Times. Retrieved 24 February 2020.

42) Ang, Hwee Min (23 January 2020). Schools step up precautions after Wuhan virus case confirmed in Singapore . CNA. Retrieved 29 January 2020.

43) Angiotensin-converting enzyme 2 (ACE2) as a SARS-CoV-2 receptor: molecular mechanisms and potential therapeutic target . Intensive Care Medicine. doi:10.1007/s00134-020-05985-9.

44) Antiviral Drug Combo Ineffective Vs. Coronavirus . WebMD. 20 March 2020.

45) Archives, The National. Great Plague of 1665–1666 – The National Archives . Archived from the original on 28 April 2016. Retrieved 4 July 2016.

46) As Chinese Internet Users Try To Track The Coronavirus, Their Government Is Tracking Them . BuzzFeed News. Retrieved 8 February 2020.

47) As deadly coronavirus spreads, U.S. to expand screening of passengers from China at 20 airports . Washington Post. 27 January 2020. Archived from the original on 28 January 2020.

48) Aspinall, R., & Lang, P. O. (2018). Interleukin-7 and Immunorejuvenation In: Fulop T., Franceschi C., Hirokawa K., Pawelec G. (eds) Handbook of Immunosenescence. Springer, Cham. https://doi.org/10.1007/978-3-319-64597-1_72-1

49) Aspinall, R., & Lang, P. O. (2018). Interventions to restore appropriate immune function in the elderly. Immunity & Ageing, 15(1), 5. doi:10.1186/s12979-017-0111-6 PMC 5785902

50) At least 65 more people confirmed with coronavirus on cruise ship in Japan . The Straits Times. 10 February 2020. Retrieved 11 February 2020.

51) At least three cases among 4,000 passengers on China-Vietnam cruise, doctor says . South China Morning Post. Archived from the original on 5 February 2020. Retrieved 3 February 2020.

52) Austin Alchon, Suzanne (2003). A pest in the land: new world epidemics in a global perspective. University of New Mexico Press. p. 21. ISBN 978-0-8263-2871-7. Archived from the original on 2019-04-01. Retrieved 2016-04-22.

53) Australia to close schools, cancel sports in event of coronavirus outbreak | National Post . 26 February 2020.

54) Australian Health Protection Principal Committee (AHPPC) novel coronavirus statement on 1 February 2020 . Department of Health (Australia). Retrieved 1 February 2020.

55) Australian Medical Pioneers Index (AMPI) – Colonial Medical Life . Archived from the original on 29 June 2017. Retrieved 16 November 2014.

56) Austria reports first two cases of coronavirus . The Guardian. AP. 25 February 2020.

57) Azioni intraprese per favorire la ricerca e l'accesso ai nuovi farmaci per il trattamento del COVID-19 . aifa.gov.it (in Italian). Retrieved 18 March 2020.

58) B.C. confirms seventh case of COVID-19, bringing total number of Canadian cases to 11 . National Post. 24 February 2020. Retrieved 25 February 2020.

59) Baharudin, Hariz (24 January 2020). Wuhan virus: ICA implements temperature screening at Woodlands and Tuas checkpoints . The Straits Times. Retrieved 29 January 2020.

60) Bahrain announces 17 cases of new coronavirus infection . Gulf News. 25 February 2020.

61) Bahrain reports six new cases of coronavirus . Al Arabiya. 28 February 2020. Retrieved 28 February 2020.

62) Bahrain reports two new cases of coronavirus, one a Saudi national . Reuters. 28 February 2020. Retrieved 28 February2020.

63) baijiahao.baidu.com (in Chinese). Hubei Provincial New Coronavirus Infectious Pneumonia Prevention and Control Headquarters. 13 February 2020. Retrieved 17 February 2020.

64) Bande F, Arshad SS, Bejo MH, Moeini H, Omar AR (2015). Progress and challenges toward the development of vaccines against avian infectious bronchitis . Journal of Immunology Research. 2015: 424860. doi:10.1155/2015/424860. PMC 4411447.

65) Barclay E (10 March 2020). How canceled events and self-quarantines save lives, in one chart . Vox.

66) Baron, Sophie Alexandra; Devaux, Christian; Colson, Philippe; Raoult, Didier; Rolain, Jean-Marc (13 March 2020). Teicoplanin: an alternative drug for the treatment of coronavirus COVID-19? . International Journal of Antimicrobial Agents: 105944. doi:10.1016/j.ijantimicag.2020.105944. ISSN 1872-7913.

67) Barry Leadbeater. South Australian History Timeline (19th Century) . Archived from the original on 28 November 2014. Retrieved 16 November 2014.

68) Baud D, Qi X, Nielsen-Saines K, Musso D, Pomar L, Favre G (March 2020). Real estimates of mortality following COVID-19 infection . The Lancet Infectious Diseases. doi:10.1016/S1473-3099(20)30195-X.

69) Beeching NJ, Fletcher TE, Fowler R (2020). BMJ Best Practices: COVID-19 (PDF). BMJ.

70) Beijing orders quarantine for foreign arrivals from virus-hit areas . CNA. 26 February 2020. Retrieved 26 February 2020.

71) Beijing to halt all inter-province shuttle buses from Jan 26 -local media . finance.yahoo.com. Archived from the original on 25 January 2020. Retrieved 25 January 2020.

72) Beijing to postpone reopening of city's kindergartens, schools, universities: state media . Reuters. 26 January 2020. Archivedfrom the original on 27 January 2020. Retrieved 27 January 2020.

73) Belarus announces first case of coronavirus - TASS . Reuters. 28 February 2020. Retrieved 28 February 2020.

74) Bell, Walter George (1951). Belinda Hollyer (ed.). The great Plague in London (folio society ed.). Folio society by arrangement with Random House. Pages 3–5

75) Berger K (12 March 2020). The Man Who Saw the Pandemic Coming . Nautilus. Retrieved 16 March 2020.

76) Berlinger, Joshua; Seo, Yoonjung (7 February 2020). All of its neighbors have it, so why hasn't North Korea reported any coronavirus cases? . CNN. Archived from the original on 7 February 2020. Retrieved 7 February 2020.

77) Berlinger, Joshua; Seo, Yoonjung (7 February 2020). All of its neighbors have it, so why hasn't North Korea reported any coronavirus cases? . CNN. Archived from the original on 7 February 2020. Retrieved 7 February 2020.

78) Bernheim A, Mei X, Huang M, Yang Y, Fayad ZA, Zhang N, et al. (February 2020). Chest CT Findings in Coronavirus Disease-19 (COVID-19): Relationship to Duration of Infection . Radiology: 200463. doi:10.1148/radiol.2020200463.

79) Bertherat, Eric; Bekhoucha, Souad; Chougrani, Saada; Razik, Fathia; Duchemin, Jean B.; Houti, Leila; Deharib, Larbi; Fayolle, Corinne; Makrerougrass, Banaouda; Dali-Yahia, Radia; Bellal, Ramdan; Belhabri, Leila; Chaieb, Amina; Tikhomirov, Evgueni; Carniel, Elisabeth (2007). Plague Reappearance in Algeria after 50 Years, 2003 . Emerging Infectious Diseases. 13 (10): 1459–1462. doi:10.3201/eid1310.070284.

80) Beveridge, W.I.B. Influenza, the Last Great Plague (Heinemann, London, 1977).

81) Bhat, Nishi (1 February 2020). N-95 Mask Export Ban! With corona in the backdrop, Health Ministry bans export of N-95 Mask . Archived from the original on 1 February 2020. Retrieved 1 February 2020.

82) Bidokhti MR, Tråvén M, Krishna NK, Munir M, Belák S, Alenius S, Cortey M (September 2013). Evolutionary dynamics of bovine coronaviruses: natural selection pattern of the spike gene implies adaptive evolution of the strains . The Journal of General Virology. 94 (Pt 9): 2036–49. doi:10.1099/vir.0.054940-0.

83) bjnews.com.cn. Archived from the original on 26 January 2020. Retrieved 26 January 2020.

84) Blackburne, George Hugh Spencer; Anderson, T. L. (1903). Report on the outbreak of plague at Fremantle – via Wikisource.

85) BNODesk (28 February 2020). BREAKING: Netherlands reports first case of coronavirus (Tweet) – via Twitter.

86) Bocci, Michele (25 February 2020). Coronavirus, 325 casi in Italia. Altri tre morti in Lombardia e uno in Veneto, le vittime diventano 11 (in Italian) . La Repubblica. Retrieved 26 February2020.

87) BOISE STATE SUSPENDS TRAVEL TO AND FROM CHINA IN RESPONSE TO CORONAVIRUS OUTBREAK . BoiseState.edu. Archived from the original on 5 February 2020. Retrieved 5 February 2020.

88) Boseley S (17 February 2020). Coronavirus causes mild disease in four in five patients, says WHO . The Guardian. Archivedfrom the original on 18 February 2020. Retrieved 18 February 2020.

89) Boseley, Sarah; Campbell, Denis; Murphy, Simon (6 February 2020). First British national to contract coronavirus had been in Singapore . The Guardian. ISSN 0261-3077. Archived from the original on 7 February 2020. Retrieved 7 February 2020.
90) Boy, 16, among six new confirmed coronavirus cases in Hong Kong . 26 February 2020.
91) Brasil confirma primeiro caso do novo coronavírus . Folha de S.Paulo. 25 February 2020.
92) Breaking news . Joel Mossong. 29 February 2020. Retrieved 29 February 2020.
93) Breuninger, Leslie Josephs,Kevin (31 January 2020). Trump imposes travel restrictions, mandatory quarantines over coronavirus outbreak . CNBC.
94) BRIEF-Corrected-Zhejiang Hisun Pharma gets approval for clinical trial to test flu drug Favipiravir for pneumonia caused by new coronavirus. Reuters Healthcare, 16 February 2020.
95) British nationals told to leave China over coronavirus risk . South China Morning Post. 4 February 2020. Archived from the original on 5 February 2020. Retrieved 8 February 2020.
96) Britons warned against travel to mainland China . BBC News. 28 January 2020. Retrieved 28 January 2020.
97) Brown J (2018). Influenza: The Hundred Year Hunt to Cure the Deadliest Disease in History. New York: Atria. ISBN 978-1501181245.
98) Brueck H (30 January 2020). There's only one way to know if you have the coronavirus, and it involves machines full of spit and mucus . Business Insider. Archived from the original on 1 February 2020. Retrieved 1 February 2020.
99) BSI open letter to Government on SARS-CoV-2 outbreak response | British Society for Immunology . www.immunology.org. Retrieved 15 March 2020.
100) Burj Khalifa lights up red in solidarity with China over coronavirus . The National (UAE). Archived from the original on 3 February 2020. Retrieved 3 February 2020.

101) Burke, Minyvonne. Coronavirus case in Boston is 1st in Massachusetts; 8th in the U.S. NBC News. NBC Universal. Archived from the original on 2 February 2020. Retrieved 1 February 2020.

102) Butler, Declan (October 2012). SARS veterans tackle coronavirus . Nature. 490 (7418): 20. Bibcode:2012Natur.490...20B. doi:10.1038/490020a.

103) BWF Statement on Postponement of Lingshui China Masters . bwfbadminton.com. Badminton World Federation. 1 February 2020. Retrieved 2 February 2020.

104) Ca nhiễm virus corona thứ 10 tại Việt Nam . VnExpress (in Vietnamese). 4 February 2020. Archived from the original on 4 February 2020. Retrieved 4 February 2020.

105) Ca thứ 15 nhiễm virus corona là bé gái 3 tháng tuổi ở Vĩnh Phúc (in Vietnamese) . Dân trí. 11 February 2020. Retrieved 11 February 2020.

106) Ca thứ 16 tại Việt Nam dương tính nCoV (in Vietnamese) . VnExpress. 13 February 2020. Retrieved 13 February 2020.

107) Ca thứ 8 mắc virus corona ở Việt Nam . VnExpress (in Vietnamese). 3 February 2020. Archived from the original on 3 February 2020. Retrieved 3 February 2020.

108) Cabinet announces 14-day travel restrictions from China . www.guardian.co.tt.

109) Cai, Qingxian; Yang, Minghui; Liu, Dongjing; Chen, Jun; Shu, Dan; Xia, Junxia; Liao, Xuejiao; Gu, Yuanbo; Cai, Qiue; Yang, Yang; Shen, Chenguang (18 March 2020). Experimental Treatment with Favipiravir for COVID-19: An Open-Label Control Study . Engineering. doi:10.1016/j.eng.2020.03.007. ISSN 2095-8099.

110) Cambodia suffers worst dengue epidemic, 407 dead Archived2009-11-11 at the Wayback Machine, Reuters

111) Campbell C (20 January 2020). The Wuhan Pneumonia Crisis Highlights the Danger in China's Opaque Way of Doing Things . Time. Retrieved 13 March 2020.

112) Campbell, Alan (14 February 2020). Coronavirus update: 5th case revealed in B.C. Richmond News. Retrieved 14 February2020.

113) Campbell, Denis; Perraudin, Frances (10 February 2020). Coronavirus: number of confirmed UK cases rises from four to eight . The Guardian. Retrieved 10 February 2020.
114) Can you get coronavirus twice or does it cause immunity? . The Independent. 13 March 2020. Retrieved 15 March 2020.
115) Canadians, get out of China by any means possible: foreign affairs minister . National Post. 5 February 2020. Retrieved 5 February 2020.
116) Cao B, Wang Y, Wen D, Liu W, Wang J, Fan G, et al. (March 2020). A Trial of Lopinavir-Ritonavir in Adults Hospitalized with Severe Covid-19 . The New England Journal of Medicine. doi:10.1056/NEJMoa2001282.
117) Cao, Bin; Wang, Yeming; Wen, Danning; Liu, Wen; Wang, Jingli; Fan, Guohui; Ruan, Lianguo; Song, Bin; Cai, Yanping; Wei, Ming; Li, Xingwang (18 March 2020). A Trial of Lopinavir–Ritonavir in Adults Hospitalized with Severe Covid-19 . New England Journal of Medicine. doi:10.1056/NEJMoa2001282. ISSN 0028-4793.
118) Cao, Bin; Wang, Yeming; Wen, Danning; Liu, Wen; Wang, Jingli; Fan, Guohui; Ruan, Lianguo; Song, Bin; Cai, Yanping; Wei, Ming; Li, Xingwang (18 March 2020). A Trial of Lopinavir–Ritonavir in Adults Hospitalized with Severe Covid-19 . New England Journal of Medicine. 0. doi:10.1056/NEJMoa2001282. ISSN 0028-4793.
119) Casadevall A, Pirofski LA (March 2020). The convalescent sera option for containing COVID-19 . The Journal of Clinical Investigation. doi:10.1172/JCI138003.
120) Cascella M, Rajnik M, Cuomo A, Dulebohn SC, Di Napoli R (2020). Features, Evaluation and Treatment Coronavirus (COVID-19) . StatPearls. Treasure Island (FL): StatPearls Publishing. Retrieved 18 March 2020.
121) Cascella M, Rajnik M, Cuomo A, Dulebohn SC, Di Napoli R (March 2020). Features, Evaluation and Treatment Coronavirus (COVID-19) .
122) Case-control study to assess potential risk factors related to human illness caused by the Middle East Respiratory Syndrome

Coronavirus (MERS-CoV) (PDF). World Health Organization. 2014-03-28. Retrieved 2014-04-24.

123) Casey, James (1999). Early Modern Spain: A Social History. Psychology Press. p. 37. ISBN 978-0-415-13813-0.

124) Cathay Pacific asks its 27,000 employees to take three weeks unpaid leave . South China Morning Post. 5 February 2020. Archived from the original on 6 February 2020. Retrieved 8 February 2020.

125) Cathay to cut flights to mainland China by 90 per cent amid virus woes . South China Morning Post. 4 February 2020. Archivedfrom the original on 6 February 2020. Retrieved 8 February 2020.

126) Cavanagh, Dave (2003). Severe acute respiratory syndrome vaccine development: Experiences of vaccination against avian infectious bronchitis coronavirus . Avian Pathology. 32 (6): 567–582. doi:10.1080/0307945031000l621198.

127) CDC (2019-08-02). MERS in the U.S. Centers for Disease Control and Prevention. Archived from the original on 2019-12-15. Retrieved 2019-12-10.

128) CDC (2019-08-02). MERS Transmission . Centers for Disease Control and Prevention. Archived from the original on 2019-12-07. Retrieved 2019-12-10.

129) CDC (5 February 2020). CDC Tests for 2019-nCoV . Centers for Disease Control and Prevention. Archived from the original on 14 February 2020. Retrieved 12 February 2020.

130) CDC Confirms 14th Case of 2019 Novel Coronavirus . Centers for Disease Control and Prevention (United States). 12 February 2020. Retrieved 12 February 2020.

131) Centers for Disease Control (3 February 2020). Coronavirus Disease 2019 (COVID-19): Prevention & Treatment . Archived from the original on 15 December 2019. Retrieved 10 February 2020.

132) Centers for Disease Control and Prevention (11 February 2020). What to do if you are sick with 2019 Novel Coronavirus (2019-nCoV) . Archived from the original on 14 February 2020. Retrieved 13 February 2020.

133) CEPI collaborates with the Institut Pasteur in a consortium to develop COVID-19 vaccine . Coalition for Epidemic Preparedness Innovations. 19 March 2020. Retrieved 23 March 2020.

134) CEPI welcomes UK Government's funding and highlights need for $2 billion to develop a vaccine against COVID-19 . Coalition for Epidemic Preparedness Innovations, Oslo, Norway. 6 March 2020. Retrieved 23 March 2020.

135) CGTN (2 February 2020). Beijing builds mask factory in 7 days amid widespread shortage . CGTN. CGTN. Archived from the original on 2 February 2020. Retrieved 2 February 2020.

136) Chang CK, Hou MH, Chang CF, Hsiao CD, Huang TH (March 2014). The SARS coronavirus nucleocapsid protein--forms and functions . Antiviral Research. 103: 39–50. doi:10.1016/j.antiviral.2013.12.009.

137) Chang, Ai-Lien (8 February 2020). Coronavirus: 7 new cases in Singapore, possible new cluster involving church in Paya Lebar . The Straits Times. Retrieved 8 February 2020.

138) Chang, Ai-Lien; Khalik, Salma (4 February 2020). Coronavirus: S'pore reports first cases of local transmission; 4 out of 6 new cases did not travel to China . The Straits Times. Archived from the original on 4 February 2020. Retrieved 4 February 2020.

139) Chang, Ai-Lien; Khalik, Salma; Kurohi, Rei (27 January 2020). Wuhan virus: Singapore to impose 14 days leave of absence for those in schools, healthcare and eldercare who travelled to China . The Straits Times. Retrieved 27 January 2020.

140) Chapman, Michelle. Coronavirus: Royal Caribbean cancels more China cruises, adds Hong Kong to no-board list . USA TODAY. Archived from the original on 6 February 2020. Retrieved 4 February 2020.

141) Chee, Kenny (5 February 2020). Coronavirus: Singapore confirms 4 new cases, including 6-month-old child of infected couple . The Straits Times. Retrieved 5 February 2020.

142) Chen A. China's coronavirus app could have unintended consequences . MIT Technology Review. Retrieved 7 March2020.

143) Chen H, Guo J, Wang C, Luo F, Yu X, Zhang W, et al. (February 2020). Clinical characteristics and intrauterine vertical

transmission potential of COVID-19 infection in nine pregnant women: a retrospective review of medical records . Lancet. 395 (10226): 809–15. doi:10.1016/S0140-6736(20)30360-3.

144) Chen N, Zhou M, Dong X, Qu J, Gong F, Han Y, et al. (February 2020). Epidemiological and clinical characteristics of 99 cases of 2019 novel coronavirus pneumonia in Wuhan, China: a descriptive study . Lancet. 395 (10223): 507–513. doi:10.1016/S0140-6736(20)30211-7.

145) Chen WH, Strych U, Hotez PJ, Bottazzi ME (3 March 2020). The SARS-CoV-2 Vaccine Pipeline: an Overview . Current Tropical Medicine Reports. doi:10.1007/s40475-020-00201-6.

146) Chen, Eli (5 March 2020). Wash U Scientists Are Developing A Coronavirus Vaccine . news.stlpublicradio.org. Retrieved 19 March 2020.

147) Chen, Natasha. Cruise line announces restrictions on passengers who were in China . CNN. Turner Broadcasting System, Inc. Archived from the original on 1 February 2020. Retrieved 1 February 2020.

148) Chen, Wei-ting; Low, Y.F. (20 February 2020). WUHAN VIRUS/Taiwan reports new coronavirus case with unknown origin . Focus Taiwan. Retrieved 21 February 2020.

149) Chen, Wei-ting; Yeh, Joseph (24 February 2020). Two more COVID-19 cases reported, bringing Taiwan total to 30 . Focus Taiwan. Retrieved 24 February 2020.

150) Cheng ZJ, Shan J (February 2020). 2019 Novel coronavirus: where we are and what we know . Infection. doi:10.1007/s15010-020-01401-y.

151) Cheng, Lilian; Lum, Alvin; Sun, Fiona (14 February 2020). Coronavirus: Hong Kong confirms three new cases, as hospital staff say patients violated mandatory quarantine to visit doctors . South China Morning Post. Retrieved 14 February 2020.

152) Cheung E (13 March 2020). Some recovered Covid-19 patients may have lung damage, doctors say . South China Morning Post.

153) Cheung JC, Ho LT, Cheng JV, Cham EY, Lam KN (February 2020). Staff safety during emergency airway management for

COVID-19 in Hong Kong . Lancet Respiratory Medicine. doi:10.1016/s2213-2600(20)30084-9.

154) Cheung, Elizabeth (28 January 2020). Hong Kong researchers have developed coronavirus vaccine, expert reveals . South China Morning Post. Archived from the original on 28 January 2020. Retrieved 28 January 2020.

155) Cheung, Elizabeth (5 February 2020). Coronavirus: three new cases in Hong Kong include wife and daughter of man who already has disease . South China Morning Post. Archived from the original on 5 February 2020. Retrieved 5 February 2020.

156) Cheung, Elizabeth; Lum, Alvin (19 February 2020). Coronavirus: confirmed Hong Kong cases now 65 as mother-in-law of infected engineer becomes one of three more struck down in virus outbreak . South China Morning Post. Retrieved 19 February2020.

157) Cheung, Tony; Ting, Victor; Low, Zoe; Zhang, Karen; Leung, Kanis (22 February 2020). Coronavirus: as Diamond Princess cruise ship evacuees return from Japan, Hong Kong confirms 70th infection . South China Morning Post. Retrieved 23 February 2020.

158) China - Traveler view | Travelers' Health | CDC. wwwnc.cdc.gov.

159) China allots nearly $9 billion to contain spread of virus. Reuters. 27 January 2020. Archived from the original on 27 January 2020. Retrieved 27 January 2020.

160) China allows non-residents of Wuhan to leave coronavirus centre . The Straits Times. 24 February 2020. Retrieved 24 February 2020.

161) China asks EU for help with supplies . CNN. CNN. 1 February 2020. Archived from the original on 2 February 2020. Retrieved 2 February 2020.

162) China bans trade, consumption of wild animals due to coronavirus . Reuters. 25 February 2020. Retrieved 28 February2020.

163) China bans wildlife trade nationwide due to Wuhan virus outbreak . CNA. Archived from the original on 26 January 2020. Retrieved 26 January 2020.

164) China 'comprehensively' bans illegal wildlife trade after COVID-19 outbreak . CNA. 24 February 2020. Retrieved 27 February 2020.

165) China confirms human-to-human transmission of Wuhan virus as WHO emergency group meets . CNA. 21 January 2020. Retrieved 21 January 2020.

166) China coronavirus outbreak: All the latest updates . Al Jazeera. Retrieved 25 February 2020.

167) China coronavirus outbreak: Hong Kong declares emergency and closes schools . www.cbsnews.com. Retrieved 26 January 2020.

168) China Inflation Accelerates on Holiday Demand and Outbreak . Bloomberg News. Bloomberg News. 10 February 2020. Retrieved 10 February 2020.

169) China inflation rises as coronavirus disrupts supply chains . France 24. France 24. 10 February 2020. Retrieved 10 February2020.

170) China launches coronavirus 'close contact' app . BBC News. 11 February 2020. Retrieved 7 March 2020.

171) China scientists want to patent Gilead drug to treat coronavirus patients . MarketWatch. Associated Press. Archived from the original on 6 February 2020. Retrieved 8 February 2020.

172) China to extend Spring Festival holiday to contain coronavirus outbreak . Xinhua News Agency. Beijing. 26 January 2020.

173) China Travel Advisory . travel.state.gov.

174) China Travel Advisory . travel.state.gov. Retrieved 27 January2020.

175) China turns Roche arthritis drug Actemra against COVID-19 in new treatment guidelines . FiercePharma. Retrieved 14 March2020.

176) China virus spread is accelerating, Xi warns . BBC News. 25 January 2020. Archived from the original on 2 February 2020. Retrieved 8 February 2020.

177) China Wants to Patent Gilead's Experimental Coronavirus Drug . 5 February 2020. Archived from the original on 6 February 2020. Retrieved 8 February 2020.

178) China's first confirmed Covid-19 case traced back to November 17 . South China Morning Post. 13 March 2020. Archived from the original on 19 March 2020.

179) China's inflation rises amid virus outbreak . The Economist Times. AFP. 10 February 2020. Retrieved 10 February 2020.

180) China's Shantou city will not ban cars, ships, people from entering, state media reports . Reuters. 26 January 2020. Archived from the original on 27 January 2020. Retrieved 8 February 2020 – via www.reuters.com.

181) China's Tangshan city suspends public transit to prevent spread of coronavirus . Reuters. 27 January 2020. [https Reuters website://web.archive.org/web/20200127234759/https://www.reuters.com/article/us-china-health-tangshan-idUSKBN1ZQ2ICArchived] Check |archive-url= value (help) from the original on 27 January 2020. Retrieved 27 January 2020.

182) China's Tencent asks staff to work from home until February 7 due to virus . Reuters. 27 January 2020. Archived from the original on 27 January 2020. Retrieved 27 January 2020.

183) Chinese Clinical Trial Register (ChiCTR) - The world health organization international clinical trials registered organization registered platform . chictr.org.cn. Retrieved 20 March 2020.

184) Chinese doctors using plasma therapy on coronavirus, WHO says 'very valid' approach . Reuters. 17 February 2020 – via www.reuters.com.

185) Chinese Premier Li visits Wuhan, epicenter of virus outbreak . Reuters. Beijing. 27 January 2020. Archived from the original on 27 January 2020. Retrieved 8 February 2020.

186) Chinese Woman Confirmed to Be S. Korea's 28th Coronavirus Case . KBS World. 11 February 2020. Retrieved 11 February2020.

187) Chisholm, Hugh, ed. (1911). Plague . Encyclopædia Britannica. 21 (11th ed.). Cambridge University Press. pp. 693–705.

188) Chisholm, Hugh, ed. (1911). Yellow Fever . Encyclopædia Britannica. 28 (11th ed.). Cambridge University Press. pp. 910–911.

189) Cholera Country Profile: Zimbabwe (PDF). World Health Organization – Global Task Force on Cholera Control. 31 October 2009. Retrieved 16 February 2020.

190) Cholera death toll in India rises Archived 2017-11-11 at the Wayback Machine, BBC News

191) Cholera epidemic in western Chad kills 123 . 2004-09-02. Archived from the original on 2009-01-12. Retrieved 2008-06-08.
192) Cholera epidemic losing its sting Archived 2008-06-26 at the Wayback Machine
193) Cholera outbreak in Iraq growing Archived 2017-02-04 at the Wayback Machine, Associated Press
194) Cholera Spreads Through South Africa Townships . Archivedfrom the original on 3 June 2009. Retrieved 16 November 2014.
195) Chong, Clara (7 February 2020). Coronavirus: 3 new Singapore cases include Victoria JC teacher . The Straits Times. Retrieved 7 February 2020.
196) CHP investigates additional case of novel coronavirus infection . Centre for Health Protection (Hong Kong). 12 February 2020. Retrieved 12 February 2020.
197) CHP investigates three additional cases of novel coronavirus infection . Centre for Health Protection (Hong Kong). 13 February 2020. Retrieved 13 February 2020.
198) Chughtai A, Malik A (March 2020). Is Coronavirus disease (COVID-19) case fatality ratio underestimated? . Global Biosecurity. 1 (3). doi:10.31646/gbio.56 (inactive 19 March 2020).
199) Chumko, Andre (28 February 2020). Coronavirus: First case of virus in New Zealand . Stuff.co.nz. Retrieved 28 February 2020.
200) CityNews. Wuhan viral pneumonia alert . Chiang Mai Citylife. Retrieved 15 January 2020.
201) Clinical Development Success Rates 2006-2015 (PDF). BIO Industry Analysis. June 2016.
202) Clinical management of severe acute respiratory infection when novel coronavirus (nCoV) infection is suspected . World Health Organization (WHO). Archived from the original on 31 January 2020. Retrieved 13 February 2020.
203) CMO confirms fourth case of coronavirus in England . GOV.UK.
204) CMO for England announces 2 new cases of novel coronavirus . gov.uk. Retrieved 27 February 2020.

205) Cohen J (2020-01-26). Wuhan seafood market may not be source of novel virus spreading globally . ScienceMag American Association for the Advancement of Science. (AAAS). Archivedfrom the original on 2020-01-27. Retrieved 2020-01-29.
206) Cohen J, Normile D (January 2020). New SARS-like virus in China triggers alarm (PDF). Science. 367 (6475): 234–35. Bibcode:2020Sci...367..234C. doi:10.1126/science.367.6475.234. Archived (PDF) from the original on 11 February 2020. Retrieved 11 February 2020.
207) Cohen, Elizabeth; Andone, Dakin; Tinker, Ben. US government declares the novel coronavirus a public health emergency . CNN.
208) Cohen, Jon (26 January 2020). Wuhan seafood market may not be source of novel virus spreading globally . Science | AAAS. Archived from the original on 27 January 2020. Retrieved 8 February 2020.
209) Colorado researchers are racing to find an antiviral drug that could save people with the new coronavirus . 20 March 2020.
210) Companies in Shanghai cannot resume work before February 9 due to virus fight: government . Reuters. 27 January 2020. Archived from the original on 27 January 2020. Retrieved 27 January 2020.
211) Compulsory leave of absence for students, teachers returning from China: MOE . The Straits Times. 27 January 2020. Retrieved 16 February 2020.
212) Confirmed coronavirus cases in Hong Kong increase to 84 . The New Indian Express. 25 February 2020. Retrieved 26 February2020.
213) Confirmed COVID-19 cases in Austria rise to 10 . Xinhua. Retrieved 29 February 2020.
214) Confirmed Covid-19 cases increase to 74 in Hong Kong . The Star. 23 February 2020. Retrieved 24 February 2020.
215) Connelly, Eileen AJ (22 February 2020). First coronavirus death in Italy forces 10 towns into lockdown .
216) Connelly, Eileen AJ (28 February 2020). Two new coronavirus cases identified in Washington state . New York Post. Retrieved 29 February 2020.

217) Connelly, Eileen AJ (29 February 2020). Two new coronavirus cases identified in Washington state . New York Post. Retrieved 29 February 2020.

218) Controlling the global obesity epidemic, the World Health Organisation

219) Cook, Noble David; Lovell, William George (2001). Secret Judgments of God: Old World Disease in Colonial Spanish America. University of Oklahoma Press. p. 127. ISBN 978-0-8061-3377-5.

220) Corman VM, Muth D, Niemeyer D, Drosten C (2018). Hosts and Sources of Endemic Human Coronaviruses . Advances in Virus Research. 100: 163–88. doi:10.1016/bs.aivir.2018.01.001. ISBN 978-0-12-815201-0.

221) Coronavirus (COVID-19) Mortality Rate . www.worldometers.info. 5 March 2020. Retrieved 23 March 2020.

222) Coronavirus (COVID-19) Update: FDA Issues first Emergency Use Authorization for Point of Care Diagnostic (Press release). FDA. 21 March 2020. Retrieved 22 March 2020.

223) Coronavirus . who.int. Archived from the original on 30 January 2020. Retrieved 27 January 2020.

224) Coronavirus . www.who.int. Retrieved 27 January 2020.

225) Coronavirus : Air France suspend ses vols à destination et en provenance de la Chine jusqu'au 9 février . LaProvence.com (in French). 30 January 2020. Retrieved 30 January 2020.

226) Coronavirus Age, Sex, Demographics (COVID-19) . www.worldometers.info. Archived from the original on 27 February 2020. Retrieved 26 February 2020.

227) Coronavirus cases spreading in Switzerland . Swissinfo. 29 February 2020. Retrieved 29 February 2020.

228) Coronavirus claims first life outside China as Wuhan enforces quarantine . South China Morning Post. 2 February 2020. Archived from the original on 6 February 2020. Retrieved 8 February 2020.

229) Coronavirus COVID-19 Global Cases by Johns Hopkins CSSE . ArcGIS. Johns Hopkins CSSE. Retrieved 25 March 2020.

230) Coronavirus death toll in Iran rises to 34, 388 people infected: Health Ministry . Al Arabiya. 28 February 2020. Retrieved 28 February 2020.

231) Coronavirus death toll in Italy rises to 17 as 650 cases confirmed . The Local.it. Retrieved 27 February 2020.

232) Coronavirus Directo | Un nuevo positivo de coronavirus en Segovia y seis más en Valencia elevan a 22 los casos activos en España . rtve. Retrieved 27 February 2020.

233) Coronavirus Disease 2019 (COVID-19) - Transmission . Centers for Disease Control and Prevention. 17 March 2020. Retrieved 23 March 2020.

234) Coronavirus Disease 2019 (COVID-19) . Centers for Disease Control and Prevention (CDC). 15 February 2020. Archived from the original on 26 February 2020. Retrieved 20 February 2020.

235) Coronavirus Disease 2019 (COVID-19) . Centers for Disease Control and Prevention. 11 February 2020. Retrieved 8 March 2020.

236) Coronavirus Disease 2019 (COVID-19) . Centers for Disease Control and Prevention. 11 February 2020. Retrieved 11 March2020.

237) Coronavirus Disease 2019 (COVID-19) . Centers for Disease Control and Prevention. 11 February 2020. Retrieved 2 March 2020.

238) Coronavirus Disease 2019 (COVID-19) Frequently Asked Questions . Food and Drug Administration.

239) Coronavirus Disease 2019 (COVID-19) Symptoms . Centers for Disease Control and Prevention. United States. 10 February 2020. Archived from the original on 30 January 2020.

240) Coronavirus disease 2019 (COVID-19): situation report, 62 (PDF). World Health Organization. 22 March 2020. Retrieved 23 March 2020.

241) Coronavirus Disease 2019 (COVID-19)—Prevention & Treatment . Centers for Disease Control and Prevention. U.S. Department of Health & Human Services. 10 March 2020.

242) Coronavirus disease 2019 (COVID-19)Situation Report –24 (PDF). who.int. Retrieved 14 February 2020.

243) Coronavirus Disease 2019 (COVID-19)—Symptoms . Centers for Disease Control and Prevention. 20 March 2020. Retrieved 21 March 2020.

244) Coronavirus disease named Covid-19 . BBC. 11 February 2020. Retrieved 11 February 2020.

245) Coronavirus epicentre Wuhan revokes announcement easing lockdown . The Straits Times. 24 February 2020. Retrieved 24 February 2020.

246) Coronavirus erreicht Wien: 72-Jähriger erkrankt . Kurier. Retrieved 27 February 2020.

247) Coronavirus fears spread across the world . CNN. Retrieved 26 February 2020.

248) Coronavirus in Italia: 152 i contagiati, tre decessi. Stop al Carnevale di Venezia, scuole chiuse al Nord . ANSA. 23 February 2020. Retrieved 23 February 2020.

249) Coronavirus in Italia: aggiornamento ora per ora . la Repubblica. 22 February 2020. Retrieved 26 February 2020.

250) Coronavirus in Italia: gli ultimi aggiornamenti (in Italian) . la Repuubblica. 22 February 2020. Retrieved 29 February 2020.

251) Coronavirus in Norway: The Latest News on the COVID-19 Outbreak . Life in Norway. Retrieved 27 February 2020.

252) Coronavirus kills 12 in Iran, 61 infected- health ministry . National Post. 24 February 2020. Retrieved 25 February 2020.

253) Coronavirus latest: Germany's COVID-19 cases almost double . Deutsche Welle. Retrieved 1 March 2020.

254) Coronavirus Live Updates: Vietnam and Australia Curb Access to China as Death Toll Passes 300 . Archived from the original on 2 February 2020. Retrieved 2 February 2020.

255) Coronavirus llega a México; confirman primer caso en CdMx . Mileneo 2020. 28 February 2020. Retrieved 28 February 2020.

256) Coronavirus news . The Independent. Retrieved 25 February2020.

257) Coronavirus outbreak . Twitter. Dawn. 29 February 2020. Retrieved 29 February 2020.

258) Coronavirus outbreak a global health emergency, WHO declares . South China Morning Post. 31 January 2020.

259) Coronavirus outbreak: UAE Tour cancelled due to two 'suspected' cases . BBC Sport. 28 February 2020. Retrieved 28 February2020.

260) Coronavirus passed on within UK for first time as 20th case confirmed . The Northern Echo. 28 February 2020. Retrieved 28 February 2020.

261) Coronavirus spreads to more countries: All the latest updates . www.aljazeera.com.

262) Coronavirus spreads to Wisconsin as hundreds more Americans return to U.S. from outbreak's epicenter . CBS News. Archivedfrom the original on 6 February 2020. Retrieved 6 February 2020.

263) Coronavirus to be known as NCP for now, as China reports 86 new deaths . South China Morning Post. 8 February 2020. Retrieved 8 February 2020.

264) Coronavirus travel ban now in effect . www.guardian.co.tt.

265) Coronavirus Update live): 191,127 Cases and 7,807 Deaths 78,328 Recovered from 2019 – Ncov Virus Outbreak – Worldometer . www.worldometers.info. Retrieved 2020-03-16.

266) Coronavirus updates: The latest news on the outbreak and the global response . NBC News. Archived from the original on 6 February 2020. Retrieved 8 February 2020.

267) Coronavirus updates: US criticizes Iran and China over the their handling of the outbreak . Deutsche Welle. 25 February 2020.

268) Coronavirus, 374 contagiati in Italia, 12 vittime: la prima in Emilia-Romagna. LIVE (in Italian) . Sky TG24. 26 February 2020. Retrieved 26 February 2020.

269) Coronavirus, il Veneto sperimenta l'antivirale giapponese Favipiravir. Ma l'Aifa: Ci sono scarse evidenze scientifiche su efficacia . Il Fatto Quotidiano (in Italian). 22 March 2020. Retrieved 23 March 2020.

270) Coronavirus, Oms: 'Livello minaccia mondiale virus ora molto alta' - Salute & Benessere . ANSA.it. 22 February 2020.

271) Coronavirus, salgono a sei i contagi: i primi sono un 38enne di Codogno, sua moglie e un amico. In isolamento 250 persone . La Repubblica. 21 February 2020. Retrieved 21 February 2020.

272) Coronavirus, últimas noticias en directo: Dos nuevos casos positivos en Madrid suben la cifra a catorce . ABC Sociedad. Retrieved 27 February 2020.

273) Coronavirus, via libera dell'Aifa al farmaco anti-artrite efficace su 3 pazienti e a un antivirale: test in 5 centri [Coronavirus, Aifa gives go-ahead to effective anti-arthritis drug on 3 patients and an antiviral: test in 5 centers]. Il Messaggero (in Italian). Retrieved 14 March 2020.

274) Coronavirus: 10 passengers stuck on Diamond Princess cruise ship infected . South China Morning Post. 5 February 2020. Archived from the original on 6 February 2020. Retrieved 8 February 2020.

275) Coronavirus: 16 new cases in Italy . ANSA. 21 February 2020. Retrieved 21 February 2020.

276) Coronavirus: 20 infected on quarantined cruise ship in Japan . South China Morning Post. 6 February 2020. Archived from the original on 7 February 2020. Retrieved 8 February 2020.

277) Coronavirus: 229 infected in Italy - English . ANSA.it. 24 February 2020.

278) Coronavirus: 3 children among new cases | NHK WORLD-JAPAN News . NHK WORLD.

279) Coronavirus: 44 more cases found on Japan cruise ship . The Straits Times. 13 February 2020. Retrieved 13 February 2020.

280) Coronavirus: Australia to bar foreign nationals arriving from Iran . South China Morning Post. 29 February 2020. Retrieved 29 February 2020.

281) Coronavírus: China pede ajuda à UE para facilitar venda de material médico . Youtube. euronews (em português). Retrieved 2 February 2020.

282) Coronavirus: China reports 47 more deaths, 427 new cases . The Straits Times. 29 February 2020. Retrieved 29 February 2020.

283) Coronavirus: Chinese military takes control of medical supplies in Wuhan . South China Morning Post. 3 February 2020. Archivedfrom the original on 5 February 2020. Retrieved 8 February 2020.

284) Coronavirus: Chinese students in South Korea advised to take leave of absence, as cases rise . South China Morning Post. 17 February 2020. Retrieved 17 February 2020.

285) Coronavirus: Commission offers financing to innovative vaccines company CureVac . European Commission. 16 March 2020. Retrieved 19 March 2020.

286) Coronavirus: Delta suspends US-China flights, pilots sue American Airlines . South China Morning Post. 31 January 2020.
287) Coronavirus: Europe on alert as four more deaths reported in Italy – updates . The Guardian. Retrieved 25 February 2020.
288) Coronavirus: First British coronavirus death as man from Diamond Princess cruise ship dies . Sky News. 28 February 2020. Retrieved 28 February 2020.
289) Coronavirus: First case confirmed in Wales . BBC. 28 February 2020. Retrieved 28 February 2020.
290) Coronavirus: First death confirmed in Europe . BBC. 15 February 2020. Retrieved 15 February 2020.
291) Coronavirus: First death confirmed in Japan | NHK WORLD-JAPAN News . NHK WORLD. Retrieved 13 February 2020.
292) Coronavirus: Five Brits in French ski chalet catch virus . BBC. 8 February 2020. Retrieved 8 February 2020.
293) Coronavirus: Four more dead in Italy as disease spreads across Europe . Sky News. Retrieved 25 February 2020.
294) Coronavirus: Four more people diagnosed in UK . BBC.
295) Coronavirus: Georgia Suspends Flights with China . Georgia Today. Retrieved 30 January 2020.
296) Coronavirus: How CID officers figured out link between 3 seemingly unrelated cases in just 1 day . Mothership.sg. Retrieved 12 February 2020.
297) Coronavirus: India Issues Fresh Travel Advisory for China . The Weather Channel. Archived from the original on 6 February 2020. Retrieved 5 February 2020.
298) Coronavirus: Iraq, Oman confirm first cases, halt flights to Iran . The Straits Times. 24 February 2020. Retrieved 24 February 2020.
299) Coronavirus: Italian national in Lombardy region tests positive . The Straits Times. 21 February 2020. Retrieved 21 February 2020.
300) Coronavirus: Italy reports first deaths, South Korea sees cases double . DW. 22 February 2020. Retrieved 22 February 2020.
301) Coronavirus: Japan reports more cases outside ship | NHK WORLD-JAPAN News . NHK WORLD.

302) Coronavirus: la barre des 100 cas atteinte en France, un premier mort aux États-Unis (in French) . BFM TV. 29 February 2020. Retrieved 29 February 2020.

303) Coronavirus: le maire de La Balme-de-Sillingy, en Haute-Savoie, annonce 2 nouveaux cas . BFM. 28 February 2020. Retrieved 28 February 2020.

304) Coronavirus: le nombre de cas de contamination confirmés en France passe de 18 à 38 . BFM.TV. Retrieved 27 February 2020.

305) Coronavirus: Malaysia confirms first case with travel history to Japan . The Straits Times. 27 February 2020. Retrieved 27 February 2020.

306) Coronavirus: Passengers disembark from Diamond Princess in Japan . BBC. 19 February 2020. Retrieved 19 February 2020.

307) Coronavirus: Scientists could repurpose drugs to treat infection . Medical News Today. Retrieved 20 March 2020.

308) Coronavirus: Singapore closes borders to all foreign travellers from China . South China Morning Post. 31 January 2020.

309) Coronavirus: Singapore confirms 2 new local cases; 1 appears to have no link to previous cases . The Straits Times. 6 February 2020. Archived from the original on 6 February 2020. Retrieved 6 February 2020.

310) Coronavirus: Singapore ups outbreak alert to Orange as more cases surface with no known links; more measures in force . The Straits Times. 7 February 2020.

311) Coronavirus: South Korea confirms 70 more coronavirus cases, total 833; 7th death reported . The Straits Times. 24 February 2020. Retrieved 24 February 2020.

312) Coronavirus: South Korea declares highest alert as infections surge . BBC. 23 February 2020. Retrieved 23 February 2020.

313) Coronavirus: South Korea raises alert to highest level as number of cases jumps to 602, with 5 deaths . The Straits Times. 23 February 2020. Retrieved 23 February 2020.

314) Coronavirus: South Korean woman travelling home from Thailand tests positive for virus . The Straits Times. 4 February 2020. Archived from the original on 4 February 2020. Retrieved 4 February 2020.

315) Coronavirus: South Korea's total cases jump to 3,150 as it reports first case of reinfection . The Straits Times. 29 February 2020. Retrieved 1 March 2020.

316) Corona-Virus: Suche nach weiteren Infizierten . WDR. Retrieved 27 February 2020.

317) Coronavirus: Taiwan reports 2 more cases, tally rises to 26 . The Straits Times. 21 February 2020. Retrieved 21 February 2020.

318) Coronavirus: Taiwan reports five new cases; total at 39 . The Straits Times. 29 February 2020. Retrieved 29 February 2020.

319) Coronavirus: Tenerife hotel with hundreds of guests locked down . BBC News. 25 February 2020. Retrieved 25 February2020.

320) Coronavirus: Thailand records one new case, bringing total to 35 . The Straits Times. 17 February 2020. Retrieved 17 February2020.

321) Coronavirus: Three more people test positive in England . BBC News. 29 February 2020. Retrieved 29 February 2020.

322) Coronavirus: Two Malaysians flown back from Wuhan test positive, bringing total to 12 . The Straits Times. 5 February 2020. Archived from the original on 6 February 2020. Retrieved 8 February 2020.

323) Coronavirus: two more women infected after visiting Hong Kong temple at centre of Covid-19 cluster . South China Morning Posr. Retrieved 27 February 2020.

324) Coronavirus: Two passengers dead from quarantined Diamond Princess . BBC. 20 February 2020. Retrieved 20 February 2020.

325) Coronavirus: un nouveau cas confirmé en France, la femme d'un patient hospitalisé à Annecy . BFM.TV. Retrieved 26 February2020.

326) Coronavirus: United Airlines suspends US flights to and from Hong Kong . South China Morning Post. 5 February 2020. Archived from the original on 6 February 2020. Retrieved 8 February 2020.

327) Coronavirus: WHO enlists internet giants in war on rumours . South China Morning Post. 3 February 2020. Archived from the original on 5 February 2020. Retrieved 8 February 2020.

328) Covid-19 - Situazione in Italia (in Italian) . Ministry of Health (Italy). 28 February 2020. Retrieved 28 February 2020.

329) COVID-19 in Canada: nation's 13th case found in Toronto as repatriated Canadians released from quarantine . The Georgia Straight. 27 February 2020.

330) COVID-19 updates: Germany facing a 'coronavirus epidemic' . DW.com. Retrieved 26 February 2020.

331) Covid-19: Detetado em Londres o primeiro caso do novo coronavírus . lusa.pt. Retrieved 12 February 2020.

332) COVID-19: Malaysia bars travellers from South Korea, confirms 2 new cases . CNA. 28 February 2020. Retrieved 28 February 2020.

333) COVID-19: mRNA vaccines – a promising approach to vaccine development . Shelston IP Australia – Intellectual Property & Patent Services | IP Attorneys & IP Lawyers. 23 March 2020. Retrieved 23 March 2020.

334) COVID19—Resources for Health Care Professionals . Penn Libraries. 11 March 2020. Retrieved 13 March 2020.

335) Croatia latest European country to confirm coronavirus case . The Guardian. 25 February 2020.

336) Crossley BM, Mock RE, Callison SA, Hietala SK (December 2012). Identification and characterization of a novel alpaca respiratory coronavirus most closely related to the human coronavirus 229E . Viruses. 4 (12): 3689–700. doi:10.3390/v4123689. PMC 3528286.

337) Cruise ships to ban recent China visitors: Global industry bodyArchived 5 February 2020 at the Wayback Machine in Al Arabiya English. 3 February 2020.

338) Cruz JL, Becares M, Sola I, Oliveros JC, Enjuanes L, Zúñiga S (September 2013). Alphacoronavirus protein 7 modulates host innate immune response . Journal of Virology. 87 (17): 9754–67. doi:10.1128/JVI.01032-13.

339) Cruz JL, Sola I, Becares M, Alberca B, Plana J, Enjuanes L, Zuñiga S (June 2011). Coronavirus gene 7 counteracts host defenses and modulates virus virulence . PLOS Pathogens. 7 (6): e1002090. doi:10.1371/journal.ppat.1002090.

340) CT provides best diagnosis for COVID-19 . ScienceDaily. 26 February 2020. Retrieved 2 March 2020.

341) Cui J, Han N, Streicker D, Li G, Tang X, Shi Z, et al. (October 2007). Evolutionary relationships between bat coronaviruses and their hosts . Emerging Infectious Diseases. 13 (10): 1526–32. doi:10.3201/eid1310.070448. PMC 2851503.

342) Cui J, Li F, Shi ZL (March 2019). Origin and evolution of pathogenic coronaviruses . Nature Reviews. Microbiology. 17 (3): 181–192. doi:10.1038/s41579-018-0118-9.

343) Different SARS-CoV strains isolated from several hosts vary in their binding affinities for human ACE2 and consequently in their infectivity of human cells76,78.

344) Cui P, Chen Z, Wang T, Dai J, Zhang J, Ding T, Jiang J, Liu J, Zhang C, Shan W, Wang S, Rong Y, Chang J, Miao X, Ma X, Wang S (27 February 2020). Clinical features and sexual transmission potential of SARS-CoV-2 infected female patients: a descriptive study in Wuhan, China. MedRxiv (preprint). doi:10.1101/2020.02.26.20028225.

345) Culbertson, Alix (21 February 2020). Coronavirus: South Korea new COVID-19 front as 100 fresh cases confirmed . Sky News. Retrieved 21 February 2020.

346) Curetis Group Company Ares Genetics and BGI Group Collaborate to Offer Next-Generation Sequencing and PCR-based Coronavirus (2019-nCoV) Testing in Europe . GlobeNewswire News Room. 30 January 2020. Archived from the original on 31 January 2020. Retrieved 1 February 2020.

347) D. Ch. Stathakopoulos Famine and Pestilence in the late Roman and early Byzantine Empire (2007) 95

348) Daly, Walter J. (2008). The Black Cholera Comes to the Central Valley of America in the 19th Century – 1832, 1849, and Later . Transactions of the American Clinical and Climatological Association. 119: 143–153. ISSN 0065-7778.

349) Dannewitz, Juliane (23 March 2020). Hackathon Germany: #WirvsVirus . Datenschutzbeauftragter (in German).

350) Dansker smittet med coronavirus . tv2.dk. 27 February 2020. Retrieved 27 February 2020.

351) Davidson, Helen (13 March 2020). Covid-19 case happened in November, China government records show – report . The Guardian. ISSN 0261-3077. Retrieved 15 March 2020.

352) Davidson, Helen (13 March 2020). First Covid-19 case happened in November, China government records show—report . The Guardian. Retrieved 21 March 2020.

353) Day, Michael (17 March 2020). Covid-19: ibuprofen should not be used for managing symptoms, say doctors and scientists . BMJ. 368: m1086. doi:10.1136/bmj.m1086.

354) Day, Michael (17 March 2020). Covid-19: ibuprofen should not be used for managing symptoms, say doctors and scientists . BMJ. 368. doi:10.1136/bmj.m1086. ISSN 1756-1833.

355) Death of Forty Thousand Fijians from Measles . Liverpool Mercury. 29 Sep 1875. Retrieved 9 Nov 2012.

356) Death toll Gujarat . Business Standard. 15 March 2015. Archived from the original on 2 April 2015. Retrieved 15 March2015.

357) Deaths in China Rise, With No Sign of Slowdown . 4 February 2020. Archived from the original on 6 February 2020. Retrieved 8 February 2020 – via NYTimes.com.

358) Deerwester, Jayme. Not coronavirus: 6,000 cruise ship passengers given OK to disembark after passenger diagnosed with flu . USA TODAY.

359) Defense Department Press Briefing Investigating and Developing Vaccine Candidates Against COVID-19 (Transcript) . Arlington, VA: United States Department of Defense. 5 March 2020. Retrieved 19 March 2020.

360) Definition of Coronavirus by Merriam-Webster, Merriam-Webster, retrieved 24 March 2020

361) Delta to temporarily suspend all U.S.-China flying . Delta News Hub. Retrieved 22 February 2020.

362) Democratic Republic of Congo: More measles vaccinations needed . Médecins Sans Frontières (MSF) International. Archived from the original on 2 November 2014. Retrieved 16 November 2014.

363) Dengue and severe dengue . World Health Organization (WHO). 2 March 2020.

364) Dengue cases in Philippines rise by 43 percent: government . Archived from the original on 17 October 2017. Retrieved 16 November 2014.

365) Dengue epidemic threatens India's capital . News-Medical.net. 2 October 2006. Archived from the original on 15 January 2009. Retrieved 16 November 2014.

366) Dengue fever epidemic hits Caribbean, Latin AmericaArchived 2009-08-03 at the Wayback Machine, Reuters

367) Dengue fever in Indonesia – update 4 11 May 2004 www.who.int, accessed 16 February 2020

368) Dengue in the Americas: The Epidemics of 2000 . Archivedfrom the original on 2017-10-17. Retrieved 2008-08-05.

369) Desjardins, Bertrand (1996). Demographic Aspects of the 1702–1703 Smallpox Epidemic in the St. Lawrence Valley . Canadian Studies in Population. 23 (1): 49–67. doi:10.25336/P6459C.

370) Devaux, Christian A. (2013). Small oversights that led to the Great Plague of Marseille (1720–1723): Lessons from the past . Infection, Genetics and Evolution. 14: 169–185. doi:10.1016/j.meegid.2012.11.016.

371) Devlin, Hannah (24 January 2020). Lessons from SARS outbreak help in race for coronavirus vaccine . The Guardian. Archived from the original on 25 January 2020. Retrieved 25 January 2020.

372) Devlin, Hannah; Sample, Ian (10 March 2020). Hopes rise over experimental drug's effectiveness against coronavirus . The Guardian. Retrieved 19 March 2020.

373) Dhal, Sharmila (22 February 2020). Iranian couple diagnosed with COVID-19 in UAE, taking number of cases to 13 . Gulf News. Retrieved 24 February 2020.

374) Diagnosis and Treatment of New Coronavirus Infected Pneumonia (Trial Version 5)] (PDF). Archived (PDF) from the original on 5 February 2020. Retrieved 5 February 2020.

375) Diao Y, Liu X, Wang T, Zeng X, Dong C, Zhou C, Zhang Y, She X, Liu D, Hu Z (20 February 2020). Estimating the cure rate and case fatality rate of the ongoing epidemic COVID-19 . MedRxiv (preprint). doi:10.1101/2020.02.18.20024513.

376) Diffusione Coronavirus: Tencent ha pubblicato i veri dati nascosti dal governo cinese? L'analisi . Open (in Italian). 7 February 2020. Retrieved 8 February 2020.

377) Doctored Screenshots Claim China's Coronavirus Deaths Over 24,000 . The Quint. 7 February 2020. Retrieved 8 February 2020.

378) Doctorswithoutborders.org . MSF USA. Archived from the original on 27 September 2011. Retrieved 16 November 2014.

379) Dong Y, Mo X, Hu Y, Qi X, Jiang F, Jiang Z, Tong S (2020). Epidemiological Characteristics of 2143 Pediatric Patients With 2019 Coronavirus Disease in China (PDF). Pediatrics: e20200702. doi:10.1542/peds.2020-0702.

380) Donnelly, Christl A.; Malik, Mamun R.; Elkholy, Amgad; Cauchemez, Simon; Kerkhove, Maria D. Van (2019). Worldwide Reduction in MERS Cases and Deaths since 2016 – Volume 25, Number 9 – September 2019 – Emerging Infectious Diseases journal – CDC . Emerging Infectious Diseases. 25 (9): 1758–1760. doi:10.3201/eid2509.190143.

381) Doremalen N, Bushmaker T, Morris DH, Holbrook MG, Gamble A, Williamson BN, et al. (March 2020). Aerosol and Surface Stability of SARS-CoV-2 as Compared with SARS-CoV-1 . The New England Journal of Medicine. Massachusetts Medical Society. doi:10.1056/nejmc2004973.

382) Doucleef M (2012-09-26). Scientists Go Deep On Genes Of SARS-Like Virus . Associated Press. Archived from the original on 2012-09-27. Retrieved 2012-09-27.

383) Dozens of passengers allowed off virus-hit Diamond Princess ship have symptoms: Japan minister . CNA. 26 February 2020. Retrieved 26 February 2020.

384) DR Congo measles: More than 6,000 dead in world's worst outbreak . BBC News. 2020-01-08. Retrieved 2020-02-28.

385) Duarte, Fernando (24 February 2020). Who is 'patient zero' in the coronavirus outbreak? . BBC. Archived from the original on 8 March 2020.

386) Duddu P (19 February 2020). Coronavirus outbreak: Vaccines/drugs in the pipeline for Covid-19 . clinicaltrialsarena.com. Archived from the original on 19 February 2020.

387) Dừng buôn bán vùng biên để ứng phó dịch corona . tuoitre.vn(in Vietnamese). 30 January 2020. Retrieved 30 January 2020.

388) Dừng tàu khách liên vận quốc tế Việt Nam - Trung Quốc , Zing News (in Vietnamese), archived from the original on 4 February 2020, retrieved 4 February 2020

389) Dừng tiếp nhận lao động Trung Quốc , thanhnien.vn (in Vietnamese), retrieved 2 February 2020

390) Dừng toàn bộ chuyến bay giữa Việt Nam và Trung Quốc từ chiều nay . tuoitre.vn (in Vietnamese). Archived from the original on 1 February 2020. Retrieved 1 February 2020.

391) Dutt, Ashok (2006). Surat Plaque of 1994 re-examined (PDF). Southeast Asian Journal of Tropical Medicine and Public Health. 37(4): 755–60. Archived (PDF) from the original on 6 February 2016. Retrieved 19 June 2016.

392) Ebola Outbreak in West Africa (section titled 'Impact') . www.cdc.gov. 2019-08-22. Retrieved 2020-02-14.

393) Ebola virus disease (Press release). World Health Organization (WHO). 12 February 2018. Retrieved 26 February2020.

394) Ebola Virus Disease Outbreak Uganda Situation Reports . WHO | Regional Office for Africa. Archived from the original on 20 June 2019. Retrieved 3 September 2019.

395) Echenberg, Myron (2007). Plague Ports: The Global Urban Impact of Bubonic Plague: 1894–1901. Sacramento: New York University Press. p. 231. ISBN 978-0-8147-2232-9.

396) Ecuador confirmó el primer caso de coronavirus (in Spanish) . Infobae. Retrieved 29 February 2020.

397) Edwards, Erika (10 February 2020). Coronavirus case confirmed in California is 13th in the U.S. NBC News. Retrieved 10 February 2020.

398) Eestis leiti esimene koroonaviirusesse nakatunu . EESTI. Retrieved 27 February 2020.

399) Effective Treatment of Severe COVID-19 Patients with Tocilizumab . ChinaXiv.org. 5 March 2020. doi:10.12074/202003.00026 (inactive 16 March 2020). Retrieved 14 March 2020.

400) Efficacy and Safety of Hydroxychloroquine for Treatment of Pneumonia Caused by 2019-nCoV (HC-nCoV) . clinicaltrials.gov. Retrieved 20 March 2020.

401) Efrati, Ido (23 February 2020). Second Coronavirus Case Diagnosed in Israel . Haaretz. Retrieved 24 February 2020.

402) Egypt reports first coronavirus case . Gulf News. 14 February 2020. Retrieved 14 February 2020.

403) Eight case of coronavirus confirmed in UAE . Middle East Monitor. 10 February 2020. Retrieved 10 February 2020.

404) Eight coronavirus cases confirmed in Bahrain, including four Saudi Arabian women . Al Arabiya. Retrieved 25 February 2020.

405) Ekblom, Jonas; Augustsson, Tomas (29 February 2020). Två nya coronafall i Sverige – 13 totalt i landet . Svenska Dagbladet (in Swedish). Retrieved 29 February 2020.

406) El cúmulo de contagios locales por coronavirus en Torrejón alerta a los expertos (in Spanish) . El Pais. 29 February 2020. Retrieved 1 March 2020.

407) Elderly woman the latest and 17th coronavirus case - Health DG . Bernama. 9 February 2020. Retrieved 9 February 2020.

408) Elise Mak (18 March 2020). China approves first homegrown COVID-19 vaccine to enter clinical trials . BioWorld. Retrieved 24 March 2020.

409) Emergencies Coronavirus EC Meeting 22 January 2020 (PDF). World Health Organization. Retrieved 31 January 2020.

410) Encephalitis outbreak: AES is a perennial issue in eastern Uttar Pradesh, northern Bihar Bihar's AES data summary looks more like a repeat of 2017 when a major JEV outbreak in Uttar Pradesh's Gorakhpur claimed the lives of many children. 17 June 2019 www.indiatvnews.com, accessed 17 February 2020

411) Encyclopedia.com . Archived from the original on 4 March 2016. Retrieved 16 November 2014.

412) Enteric Coronavirus . Diseases of Research Animals. Archived from the original on 2019-07-01. Retrieved 2020-01-24.

413) Epidemia COVID-19. Aggiornamento nazionale 19 marzo 2020 (PDF) (Report) (in Italian). Rome: Istituto Superiore di Sanità. 19 March 2020. Retrieved 22 March 2020.

414) Epidemic situation of new coronavirus pneumonia in Hubei Province on February 11, 2020] (in Chinese).
415) Epidemic. Black's Medical Dictionary, 42nd Edition. London: A&C Black, 201e. Web. 17 September 2012.
416) Epidemics. Archived from the original on 2013-07-22. Retrieved 2008-06-08.
417) Epidemiologic Clues to SARS Origin in China June 2004, www.ncbi.nlm.nih.gov, accessed 16 February 2020
418) Epidemiology of Dengue Disease in the Philippines (2000–2011): A Systematic Literature Review November 2014 PLoS Neglected Tropical Diseases via www.researchgate.net, accessed 16 February 2020
419) Epidemiology of Recurrent Hand, Foot and Mouth Disease, China, 2008–2015 wwwnc.cdc.gov Emerging Infectious Diseases, Volume 24, Number 3 – March 2018, Accessed 23 February 2020
420) ERR, BNS (21 March 2020). President makes global call to combat coronavirus via hackathon . ERR.
421) Erster bestätigter Fall in der Schweiz (in German). Blick. 25 February 2020.
422) Eschner K (2020-01-28). We're still not sure where the COVID-19 really came from . Popular Science. Archived from the original on 2020-01-29. Retrieved 2020-01-30.
423) Eschner K (2020-01-28). We're still not sure where the COVID-19 really came from . Popular Science. Archived from the original on 2020-01-29. Retrieved 2020-01-30.
424) euronews (em português) (4 February 2020). Bélgica confirma caso de infeção por coronavírus . Youtube (in Portuguese). euronews (em português). Retrieved 4 February 2020. Existem 24 casos confirmados de infeção por coronavirus no território da União Europeia, segundo dados disponíveis, terça-feira de manhã, quando foi confirmado um caso na Bélgica.
425) European Centre for Disease Prevention and Control
426) Everington, Keoni (17 February 2020). Taiwan reports 21st, 22nd cases of Wuhan coronavirus . Taiwan News. Retrieved 17 February 2020.

427) Ewing, Jack; Boudette, Neal E.; Abdul, Geneva (4 February 2020). Virus Exposes Cracks in Carmakers' Chinese Supply Chains . Archived from the original on 6 February 2020. Retrieved 8 February 2020 – via NYTimes.com.

428) Existing Drugs May Offer a First-Line Treatment for Coronavirus Outbreak . Retrieved 20 March 2020.

429) Facebook. Iraqi Ministry of Health.

430) Falco M (2012-09-24). New SARS-like virus poses medical mystery . CNN Health. Archived from the original on 2013-11-01. Retrieved 2013-03-16.

431) Fan Y, Zhao K, Shi ZL, Zhou P (March 2019). Bat Coronaviruses in China . Viruses. 11 (3): 210. doi:10.3390/v11030210. PMC 6466186.

432) Fan, Hua-Hao; Wang, Li-Qin; Liu, Wen-Li; An, Xiao-Ping; Liu, Zhen-Dong; He, Xiao-Qi; Song, Li-Hua; Tong, Yi-Gang (6 March 2020). Repurposing of clinically approved drugs for treatment of coronavirus disease 2019 in a 2019-novel coronavirus (2019-nCoV) related coronavirus model . Chinese Medical Journal: 1. doi:10.1097/CM9.0000000000000797. ISSN 2542-5641.

433) Fang L, Karakiulakis G, Roth M (March 2020). Are patients with hypertension and diabetes mellitus at increased risk for COVID-19 infection? . The Lancet Respiratory Medicine. 395 (10224): e40. doi:10.1016/S0140-6736(20)30311-1.

434) Farber, Madeline (24 February 2020). Afghanistan sees first coronavirus case as numbers spike in neighboring Iran . Fox News. Retrieved 24 February 2020.

435) Farkas J (March 2020). COVID-19—The Internet Book of Critical Care (digital) (Reference manual). USA: EMCrit. Archived from the original on 11 March 2020. Retrieved 13 March 2020.

436) Faruque, S. M.; Islam, M. J.; Ahmad, Q. S.; Faruque, A. S. G.; Sack, D. A.; Nair, G. B.; Mekalanos, J. J. (2005). Self-limiting nature of seasonal cholera epidemics: Role of host-mediated amplification of phage . Proceedings of the National Academy of Sciences. 102(17): 6119–6124. Bibcode:2005PNAS..102.6119F. doi:10.1073/pnas.0502069102. PMC 1087956.

437) Fauci, Anthony S.; Lane, H. Clifford; Redfield, Robert R. (28 February 2020). Covid-19 — Navigating the Uncharted . New England Journal of Medicine. doi:10.1056/nejme2002387. ISSN 0028-4793.

438) February 10 coronavirus news . CNN. 10 February 2020. Retrieved 10 February 2020.

439) Feek, Belinda (2 February 2020). Coronavirus: NZ Government to block travellers from mainland China, as details of recovery mission revealed . New Zealand Herald. Archived from the original on 2 February 2020. Retrieved 2 February 2019.

440) Fehr AR, Perlman S (2015). Maier HJ, Bickerton E, Britton P (eds.). Coronaviruses: an overview of their replication and pathogenesis . Methods in Molecular Biology. Springer. 1282: 1–23. doi:10.1007/978-1-4939-2438-7_1. ISBN 978-1-4939-2438-7.

441) Fehr AR, Perlman S (2015). Maier HJ, Bickerton E, Britton P (eds.). Coronaviruses: an overview of their replication and pathogenesis . Methods in Molecular Biology. Coronaviruses: Methods and Protocols. Springer. 1282: 1–23. doi:10.1007/978-1-4939-2438-7_1. ISBN 978-1-4939-2438-7.

442) Feuer, Berkeley Lovelace Jr ,William (5 February 2020). Coronavirus live updates: US can't keep virus 'out of our border,' NYC has 2 new 'patients under investigation' . CNBC. Archivedfrom the original on 6 February 2020. Retrieved 5 February 2020.

443) Feuer, Berkeley Lovelace Jr ,William (5 February 2020). WHO says coronavirus cases surge: 'Most cases in a single day' . CNBC. Archived from the original on 6 February 2020. Retrieved 5 February 2020.

444) Feuer, William (13 February 2020). CDC confirms 15th US coronavirus case, a Wuhan evacuee quarantined at Texas military base . CNBC.

445) Feuer, William (29 January 2020). Watch: WHO officials hold press conference on coronavirus as outbreak worsens across globe . CNBC.

446) Feuillatre, Cecile (27 February 2020). On the front lines of coronavirus with Italy's first sick mayor . CTV News. Archived from the original on 27 February 2020. Retrieved 27 February2020.

447) Filtering out Confusion: Frequently Asked Questions about Respiratory Protection, User Seal Check. The National Institute for Occupational Safety and Health (April 2018). Retrieved 16 March 2020.
448) finance.chinanews.com (in Chinese). Retrieved 10 February 2020.
449) First case of COVID-19 confirmed in Humboldt County . Humboldt County Department of Health and Human Services. 20 February 2020. Retrieved 21 February 2020.
450) First coronavirus case in Algeria . Jerusalem Post. Retrieved 25 February 2020.
451) First Europeans die from coronavirus in Italy . The Telegraph. 22 February 2020. Retrieved 22 February 2020.
452) First Global Estimates of 2009 H1N1 Pandemic Mortality Released by CDC-Led Collaboration | CDC . www.cdc.gov. 2019-11-20. Retrieved 2020-03-12.
453) Fisher D, Heymann D (February 2020). Q&A: The novel coronavirus outbreak causing COVID-19 . BMC Medicine. 18 (1): 57. doi:10.1186/s12916-020-01533-w. PMC 7047369.
454) Five foreigners with ties to Hubei province denied entry to Japan . The Asahi Shinbun. Archived from the original on 6 February 2020. Retrieved 3 February 2020.
455) Fler fall upptäckta av covid-19 (in Swedish) . Public Health Agency of Sweden. 28 February 2020. Retrieved 28 February 2020.
456) Flera nya coronafall i Sverige . SVT Nyheter. 27 February 2020. Retrieved 27 February 2020.
457) For different groups of people: how to choose masks . NHC.gov.cn. National Health Commission of the People's Republic of China. 7 February 2020. Retrieved 22 March 2020. Disposable medical masks: Recommended for: • People in crowded places • Indoor working environment with a relatively dense population • People going to medical institutions • Children in kindergarten and students at school gathering to study and do other activities
458) Foster, Harold D.; Hoffer, Abram (1 January 2007). Chapter 16 – Hyperoxidation of the Two Catecholamines, Dopamine and Adrenaline: Implications for the Etiologies and Treatment of

Encephalitis Lethargica, Parkinson's Disease, Multiple Sclerosis, Amyotrophic Lateral Sclerosis, and Schizophrenia . Oxidative Stress and Neurodegenerative Disorders. Elsevier Science B.V.: 369–382. doi:10.1016/B978-044452809-4/50157-5. Retrieved 11 February 2020.

459) Foundation, Thomson Reuters. Only three U.S. states can test for coronavirus - public lab group . news.trust.org.

460) Four Iranian MPs test positive for coronavirus . Al Arabiya. 28 February 2020. Retrieved 28 February 2020.

461) Four new UK coronavirus cases among ship evacuees . BBC. 23 February 2020. Retrieved 23 February 2020.

462) Fourth confirmed case of new coronavirus announced in Greece . amna.gr. 28 February 2020. Retrieved 28 February2020.

463) Fourth coronavirus case confirmed in Victoria, SA announces first two cases in Wuhan couple . abc.net.au. ABC News. 1 February 2020. Archived from the original on 1 February 2020. Retrieved 1 February 2020.

464) Fourth coronavirus case was through local transmission: Health Ministry source . The Daily Star Lebanon. 28 February 2020. Retrieved 28 February 2020.

465) Fourth passenger from cruise ship dies in Japan: Reports . CNA. 25 February 2020. Retrieved 25 February 2020.

466) France bans gatherings of more than 5,000 people after COVID-19 cases rise to 73 . CNA. 29 February 2020. Retrieved 29 February 2020.

467) France-Presse, Agence (15 February 2020). Chinese banks disinfect banknotes to stop spread of coronavirus . The Guardian. ISSN 0261-3077. Retrieved 15 February 2020.

468) Fyrsta tilfelli COVID-19 greint á Íslandi . Um Ruv. 28 February 2020. Retrieved 28 February 2020.

469) G7 countries seeking unified approach to coronavirus: Germany's Spahn . Reuters. 2 February 2020. Archived from the original on 3 February 2020. Retrieved 3 February 2020.

470) Gagnon, Alain; Mazan, Ryan (2009). Does exposure to infectious diseases in infancy affect old-age mortality? Evidence from

a pre-industrial population . Social Science & Medicine. 68 (9): 1609–1616. doi:10.1016/j.socscimed.2009.02.008.

471) Gallagher, Nancy. Egypt's Other Wars: Epidemics and the Politics of Public Health. Syracuse University Press, c1990. Published by the American University in Cairo Press. ISBN 977-424-295-5.

472) Gallagher, The Reverend John A. (1936). The Irish Emigration of 1847 and Its Canadian Consequences . Canadian Catholic Historical Association Report, University of Manitoba Web Site. Archived from the original on 2007-10-11. Retrieved 2008-03-23.

473) Gambrell, Jon (26 February 2020). Iran raises its death toll to 19 amid 139 coronavirus cases . Associated Press. Retrieved 26 February 2020.

474) Ganapathy, Nirmala (5 February 2020). Coronavirus: India starts screening passengers from Singapore . The Straits Times. Retrieved 10 February 2020.

475) Gao J, Tian Z, Yang X (February 2020). Breakthrough: Chloroquine phosphate has shown apparent efficacy in treatment of COVID-19 associated pneumonia in clinical studies . Bioscience Trends. 14 (1): 72–73. doi:10.5582/bst.2020.01047.

476) Gao, Wentao; Tamin, Azaibi; Soloff, Adam; d'Aiuto, Leonardo; Nwanegbo, Edward; Robbins, Paul D.; Bellini, William J.; Barratt-Boyes, Simon; Gambotto, Andrea (2003). Effects of a SARS-associated coronavirus vaccine in monkeys . The Lancet. 362(9399): 1895–1896. doi:10.1016/S0140-6736(03)14962-8.

477) Gates, Bill (28 February 2020). Responding to Covid-19 — A Once-in-a-Century Pandemic? . New England Journal of Medicine. doi:10.1056/nejmp2003762. ISSN 0028-4793.

478) Gaudreault E, Gosselin J (August 2009). Leukotriene B4 potentiates CpG signaling for enhanced cytokine secretion by human leukocytes. J. Immunol. 183 (4): 2650–8. doi:10.4049/jimmunol.0804135.

479) Gaudreault E, Gosselin J (May 2008). Leukotriene B4 induces release of antimicrobial peptides in lungs of virally infected mice . J. Immunol. 180 (9): 6211–21. doi:10.4049/jimmunol.180.9.6211.

480) Gayle (now), Damien; Rourke (earlier), Alison (28 January 2020). Coronavirus: Germany confirms first human transmission in

Europe – live updates . The Guardian. ISSN 0261-3077. Retrieved 28 January 2020.

481) Gayle, Alison Rourke (now); Molly Blackall Damien; Weaver, Matthew; Murray, Jessica; Rourke (earlier), Alison; Blackall, Molly; Rourke, Alison; Murray, Jessica (31 January 2020). Virus death toll reaches 213 in China – as it happened . The Guardian – via www.theguardian.com.

482) Geller C, Varbanov M, Duval RE (November 2012). Human coronaviruses: insights into environmental resistance and its influence on the development of new antiseptic strategies . Viruses. 4 (11): 3044–68. doi:10.3390/v4113044. PMC 3509683.

483) Generex Provides Coronavirus Update: Generex Receives Contract from Chinese Partners to Develop a COVID-19 Vaccine Using Ii-Key Peptide Vaccines (Press release). Generex. 27 February 2020. Retrieved 25 March 2020.

484) Genomic epidemiology of novel coronavirus (nCoV): Showing a single strain Wuhan/IPBCAMS-WH-01/2019 . Nextstrain.

485) Georgia confirms third case of coronavirus . Agenda.ge. 29 February 2020. Retrieved 29 February 2020.

486) Georgia reports first coronavirus case in the country: health minister . Reuters. 26 February 2020. Retrieved 26 February2020.

487) Georgian News Agency. 28 February 2020. Retrieved 28 February 2020.

488) Germany is at start of coronavirus epidemic: health minister . Reuters. Retrieved 26 February 2020.

489) Gerson, Michael (30 March 2015). The next epidemic . The Greenville News. South Carolina, US.

490) Geschiedenis 24 – Polio in Staphorst Archived 2013-07-14 at the Wayback Machine. Geschiedenis24.nl (2010-11-17). Retrieved on 2014-05-12.

491) Gezinsleden corona-patiënten Amsterdam en Tilburg ook besmet (in Dutch) . nos.nl.

492) Gignoux, Etienne; Idowu, Rachel; Bawo, Luke; Hurum, Lindis; Sprecher, Armand; Bastard, Mathieu; Porten, Klaudia (2015). Use of Capture–Recapture to Estimate Underreporting of Ebola Virus

Disease, Montserrado County, Liberia . Emerging Infectious Diseases. 21 (12): 2265–2267. doi:10.3201/eid2112.150756.

493) Gilgore, Sara (10 March 2020). Novavax's coronavirus vaccine program is getting some help from Emergent BioSolutions . Washington Business Journal. Charlotte, NC: American City Business Journals.

494) Gilgore, Sara (18 March 2020). Emergent BioSolutions dives into another coronavirus vaccine effort . Washington Business Journal. Retrieved 18 March 2020.

495) Glucocorticoid Therapy for Novel CoronavirusCritically Ill Patients With Severe Acute Respiratory Failure (Steroids-SARI) . clinicaltrials.gov. Retrieved 20 March 2020.

496) Gobcan, Sanidad (26 February 2020). Sanidad confirma un primer positivo por coronavirus de una ciudadana residente en el municipio gomero de Valle Gran Rey, que estuvo en Italia entre el 4 y el 8 de febrero. En estos momentos se encuentra ingresada y en aislamiento en el Hospital General de La Gomera .

497) Goh, Timothy (2 January 2020). Travellers arriving at Changi Airport from Wuhan to undergo temperature screening after pneumonia outbreak . The Straits Times. Archived from the original on 4 January 2020. Retrieved 15 January 2020.

498) Goh, Timothy; Kaur, Karamjit (28 February 2020). New coronavirus cluster at company in Science Park II with 4 staff infected, including 2 new cases announced . The Straits Times. Retrieved 28 February 2020.

499) Goh, Timothy; Kurohi, Rei (25 February 2020). Grace Assembly coronavirus mystery solved: Antibody tests linked mega cluster to 2 Wuhan tourists via CNY party and Life Church cluster in a world-first . The Straits Times. Retrieved 25 February 2020.

500) Goh, Timothy; Tee, Zhuo (20 February 2020). Coronavirus: 1 new case in S'pore, no known link yet to previous cases; 3 patients discharged . The Straits Times. Retrieved 20 February 2020.

501) Goh, Timothy; Tee, Zhuo (21 February 2020). Coronavirus: 10 more patients discharged; 1 new case linked to woman who had dengue . The Straits Times. Retrieved 21 February 2020.

502) Goh, Yan Han (9 February 2020). 3 new coronavirus cases including 71-year-old grandfather; 4 more recovered, 6 in critical condition in ICU . The Straits Times. Retrieved 9 February 2020.

503) Goh, Yan Han; Choo, Yun Ting (12 February 2020). 300 employees vacate MBFC Tower 3 after coronavirus case confirmed at DBS Asia Central . The Straits Times. Retrieved 12 February2020.

504) Goldsmith CS, Tatti KM, Ksiazek TG, Rollin PE, Comer JA, Lee WW, et al. (February 2004). Ultrastructural characterization of SARS coronavirus . Emerging Infectious Diseases. 10 (2): 320–6. doi:10.3201/eid1002.030913.

505) Virions acquired an envelope by budding into the cisternae and formed mostly spherical, sometimes pleomorphic, particles that averaged 78 nm in diameter.

506) Gorbalenya AE (11 February 2020). Severe acute respiratory syndrome-related coronavirus—The species and its viruses, a statement of the Coronavirus Study Group . bioRxiv (preprint). doi:10.1101/2020.02.07.937862.

507) Gouilh MA, Puechmaille SJ, Gonzalez JP, Teeling E, Kittayapong P, Manuguerra JC (October 2011). SARS-Coronavirus ancestor's foot-prints in South-East Asian bat colonies and the refuge theory . Infection, Genetics and Evolution. 11 (7): 1690–702. doi:10.1016/j.meegid.2011.06.021.

508) Gov in the Time of Corona . GovInsider. 19 March 2020. Retrieved 20 March 2020.

509) Government of Canada funds 49 additional COVID-19 research projects – Details of the funded projects . Government of Canada. 23 March 2020. Retrieved 23 March 2020.

510) Great Britain. Local Government Board (1893). Further report and papers on epidemic influenza, 1889–92: with an introduction by the medical officer of the Local Government Board. Eyre. Retrieved 29 March 2011.

511) Greece confirms first coronavirus case . Reuters. 26 February 2020. Retrieved 26 February 2020.

512) Greece reports two new coronavirus cases, cancels carnival . Reuters. 27 February 2020. Retrieved 27 February2020.

513) Green MS, Swartz T, Mayshar E, Lev B, Leventhal A, Slater PE, Shemer J (2002). When is an epidemic an epidemic? (PDF). Israel Medical Association Journal. 4 (1): 3–6.

514) Green MS; Swartz T; Mayshar E; Lev B; Leventhal A; Slater PE; Shemer Js (January 2002). When is an epidemic an epidemic? . Isr. Med. Assoc. J. 4 (1): 3–6.

515) Greenough, Thomas C.; Babcock, Gregory J.; Roberts, Anjeanette; Hernandez, Hector J.; et al. (15 February 2005). Development and Characterization of a Severe Acute Respiratory Syndrome–Associated Coronavirus–Neutralizing Human Monoclonal Antibody That Provides Effective Immunoprophylaxis in Mice . The Journal of Infectious Diseases. 191 (4): 507–14. doi:10.1086/427242.

516) Greg Lange, Smallpox epidemic ravages Native Americans on the northwest coast of North America in the 1770s Archived 2008-05-26 at the Wayback Machine, 23 Jan 2003, HistoryLink.org, Online Encyclopedia of Washington State History, accessed 2 Jun 2008

517) Grenfell R, Drew T (17 February 2020). Here's Why It's Taking So Long to Develop a Vaccine for the New Coronavirus . Science Alert. Archived from the original on 28 February 2020. Retrieved 26 February 2020.

518) Grenfell, Rob; Drew, Trevor (16 March 2019). Here's Why It's Taking So Long to Develop a Vaccine for the New Coronavirus . ScienceAlert. Archived from the original on 28 February 2020. Retrieved 26 February 2020.

519) Griffing, Sean M; Gamboa, Dionicia; Udhayakumar, Venkatachalam (30 August 2013). The history of 20th century malaria control in Peru . Malaria Journal. 12: 303. doi:10.1186/1475-2875-12-303.

520) Grima, Noel (19 June 2017). The 1676 plague in Malta . The Malta Independent. Archived from the original on 12 November 2017.

521) Groot RJ, Baker SC, Baric R, Enjuanes L, Gorbalenya AE, Holmes KV, Perlman S, Poon L, Rottier PJ, Talbot PJ, Woo PC, Ziebuhr J (2011). Family Coronaviridae . In King AM, Lefkowitz E, Adams MJ, Carstens EB, International Committee on Taxonomy of Viruses, International Union of Microbiological Societies. Virology Division (eds.). Ninth Report of the International Committee on

Taxonomy of Viruses. Oxford: Elsevier. pp. 806–28. ISBN 978-0-12-384684-6.

522) Growing number of airlines cancel China flights due to coronavirus . South China Morning Post. 31 January 2020.

523) Gruppo coordinamento emergenze sanitarie: aggiornamento del 27 febbraio 2020 . RTV San Marino. 27 February 2020. Retrieved 27 February 2020.

524) Gu, Jinyang; Han, Bing; Wang, Jian (27 February 2020). COVID-19: Gastrointestinal manifestations and potential fecal-oral transmission . Gastroenterology. doi:10.1053/j.gastro.2020.02.054. ISSN 0016-5085.

525) Guan WJ, Ni ZY, Hu Y, Liang WH, Ou CQ, He JX, et al. (February 2020). Clinical Characteristics of Coronavirus Disease 2019 in China . The New England Journal of Medicine. Massachusetts Medical Society. doi:10.1056/nejmoa2002032.

526) Guidance against wearing masks for the coronavirus is wrong—you should cover your face

527) Gurwitz D (March 2020). Angiotensin receptor blockers as tentative SARS-CoV-2 therapeutics . Drug Development Research. doi:10.1002/ddr.21656.

528) H, Huang; G, Hu; C, Wang; H, Xu; X, Chen; A, Qian (February 2014). Cepharanthine, an Alkaloid From Stephania Cepharantha Hayata, Inhibits the Inflammatory Response in the RAW264.7 Cell and Mouse Models . Inflammation. 37 (1): 235–46. doi:10.1007/s10753-013-9734-8.

529) Haghdoost, Yasna (20 February 2020). Iran Reports Three Confirmed Coronavirus Cases After Two Deaths . Bloomberg. Retrieved 20 February 2020.

530) Hamming, I.; Timens, W.; Bulthuis, M. L. C.; Lely, A. T.; Navis, G. J.; Goor, H. van (2004). Tissue distribution of ACE2 protein, the functional receptor for SARS coronavirus. A first step in understanding SARS pathogenesis . The Journal of Pathology. 203(2): 631–637. doi:10.1002/path.1570. ISSN 1096-9896.

531) Harper, Kyle (1 November 2017). Solving the Mystery of an Ancient Roman Plague . The Atlantic. Retrieved 20 March 2020.

532) Harris, Kathleen (29 January 2020). Canada readying to send plane to repatriate Canadians in China affected by coronavirus outbreak . CBC News. Retrieved 29 January 2020.
533) Harrison, Charlotte (27 February 2020). Coronavirus puts drug repurposing on the fast track. Nature Biotechnology. doi:10.1038/d41587-020-00003-1.
534) Hart, Brit J.; Dyall, Julie; Postnikova, Elena; Zhou, Huanying; Kindrachuk, Jason; Johnson, Reed F.; Olinger, Gene G.; Frieman, Matthew B.; Holbrook, Michael R.; Jahrling, Peter B.; Hensley, Lisa (2014). Interferon-β and mycophenolic acid are potent inhibitors of Middle East respiratory syndrome coronavirus in cell-based assaysd . The Journal of General Virology. 95 (Pt 3): 571–577. doi:10.1099/vir.0.061911-0. ISSN 0022-1317.
535) Hassan, Hazlin (4 February 2020). Malaysia confirms first citizen with coronavirus: Man was in S'pore for meeting with China colleagues last month . The Straits Times. Archived from the original on 4 February 2020. Retrieved 4 February 2020.
536) Hay un caso confirmado de coronavirus en Torreón, dice gobernador (in Spanish) . Milenio. Retrieved 29 February 2020.
537) Hays (2005). Epidemics and pandemics: their impacts on human history. ABC-CLIO. ISBN 978-1-85109-658-9. Retrieved 29 March 2011.
538) Hays, J. N. (2005). Epidemics and pandemics their impacts on human history. Santa Barbara, Calif.: ABC-CLIO. p. 103. ISBN 978-1851096589.
539) Health Ministry confirms second case of coronavirus in Israel . Jerusalem Post. 28 February 2020. Retrieved 28 February 2020.
540) Health, P. M. N. (24 February 2020). Seventh Italian dies from coronavirus in Europe's worst flare-up | National Post .
541) Hein, Alexandria (5 February 2020). 'No known effective' treatments for coronavirus despite reports, WHO says . Fox News. Archived from the original on 6 February 2020. Retrieved 6 February 2020.
542) Henderson, D (2009). Smallpox : the death of a disease. Prometheus Books. p. 12. ISBN 978-1-61592-230-7.

543) Henderson, Donald A. (30 December 2011). The eradication of smallpox – An overview of the past, present, and future . Vaccine. 29: D8. doi:10.1016/j.vaccine.2011.06.080.

544) Heng, Melissa (15 February 2020). Coronavirus: 5 new cases confirmed in Singapore, 3 linked to Grace Assembly church . The Straits Times. Retrieved 15 February 2020.

545) Henry BM (2020). COVID-19, ECMO, and lymphopenia: a word of caution . The Lancet Respiratory Medicine. Elsevier BV. doi:10.1016/s2213-2600(20)30119-3. ISSN 2213-2600.

546) Hernandez, Salvador (29 February 2020). Officials Have Confirmed The First Death From Coronavirus In The US . BuzzFeed News. Retrieved 1 March 2020.

547) Hessen MT (27 January 2020). Novel Coronavirus Information Center: Expert guidance and commentary . Elsevier Connect. Archived from the original on 30 January 2020. Retrieved 31 January 2020.

548) Heyman, Paul; Simons, Leopold; Cochez, Christel (7 January 2014). Were the English Sweating Sickness and the Picardy Sweat Caused by Hantaviruses?. Viruses. 6 (1): 151–171. doi:10.3390/v6010151.

549) Heymann DL, Shindo N (February 2020). COVID-19: what is next for public health?. Lancet. 395 (10224): 542–45. doi:10.1016/S0140-6736(20)30374-3.

550) Heymann DL, Shindo N, et al. (WHO Scientific and Technical Advisory Group for Infectious Hazards) (February 2020). COVID-19: what is next for public health? . Lancet. Elsevier BV. 395(10224): 542–545. doi:10.1016/s0140-6736(20)30374-3.

551) History of Smallpox . CDC. 25 July 2017. Retrieved 14 December 2017.

552) Ho Chi Minh City allows K-12 students one week off to avoid coronavirus spread . tuoitrenews.vn. Archived from the original on 3 February 2020. Retrieved 2 February 2020.

553) Ho, Olivia (22 February 2020). Coronavirus: 3 new cases in Singapore, bringing total to 89; 2 discharged including 6-month-old baby . The Straits Times. Retrieved 22 February 2020.

554) Hodo, Chikafumi (9 February 2020). Japan Finds Six More Coronavirus Cases on Cruise Ship . Bloomberg. Retrieved 9 February 2020.

555) Hoffmann M, Kleine-Weber H, Krüger N, Müller M, Drosten C, Pöhlmann S (31 January 2020). The novel coronavirus 2019 (2019-nCoV) uses the SARS-coronavirus receptor ACE2 and the cellular protease TMPRSS2 for entry into target cells . bioRxiv (preprint). doi:10.1101/2020.01.31.929042.

556) Hong Kong announces further border closures over Wuhan virus , CNN, archived from the original on 3 February 2020, retrieved 3 February 2020

557) Hong Kong Disneyland and Ocean Park to be closed from Sunday to help prevent spread of virus: CCTV . Reuters. 26 January 2020. Retrieved 26 January 2020.

558) Hong Kong imposes new quarantine rules over virus . bbc.com. 7 February 2020. Archived from the original on 8 February 2020. Retrieved 8 February 2020.

559) Hong Kong reports first death from coronavirus . The Straits Times. 4 February 2020. Archived from the original on 4 February 2020. Retrieved 4 February 2020.

560) Hong Kong to deny entry to anyone from Hubei to check spread of coronavirus . South China Morning Post. 26 January 2020. Retrieved 27 January 2020.

561) Hong Kong to quarantine mainland China visitors . 5 February 2020. Archived from the original on 6 February 2020. Retrieved 8 February 2020 – via www.bbc.com.

562) Hong Kong to temporarily close some of its borders with mainland China . CNN. 29 January 2020.

563) Hospital workers battling coronavirus turn to bandannas, sports goggles and homemade face shields amid shortages . The Washington Post. 19 March 2020.

564) Houston, C. S.; Houston, S. (2000). The first smallpox epidemic on the Canadian Plains: In the fur-traders' words . The Canadian Journal of Infectious Diseases. 11 (2): 112–5. doi:10.1155/2000/782978.

565) How doctors can potentially significantly reduce the number of deaths from Covid-19 . Vox. 12 March 2020. Retrieved 14 March2020.

566) Howlin, Diego (2004). Vómito Negro, Historia de la fiebre amarilla, en Buenos Aires de 1871 Archived 2017-06-07 at the Wayback Machine, Revista Persona.

567) Hsieh, Y. H.; King, C. C.; Chen, C. W.; Ho, M. S.; Lee, J. Y.; Liu, F. C.; Wu, Y. C.; Jiunn-Shyan, Julianwu (2005). Quarantine for SARS, Taiwan . Emerging Infectious Diseases. 11 (2): 278–282. doi:10.3201/eid1102.040190.

https://gisanddata.maps.arcgis.com/apps/opsdashboard/index.html#/bda7594740fd40299423467b48e9ecf6

568) https://ncats.nih.gov/preclinical/repurpose – Repurposing generally refers to studying drugs that are already approved to treat one disease or condition to see if they are safe and effective for treating other diseases.

569) https://www.nature.com/articles/s41591-020-0822-7

570) Huang C, Wang Y, Li X, Ren L, Zhao J, Hu Y, et al. (February 2020). Clinical features of patients infected with 2019 novel coronavirus in Wuhan, China . Lancet. 395 (10223): 497–506. doi:10.1016/S0140-6736(20)30183-5.

571) Huang, Chaolin; Wang, Yeming; Li, Xingwang; Ren, Lili; Zhao, Jianping; Hu, Yi; Zhang, Li; Fan, Guohui; Xu, Jiuyang; Gu, Xiaoying; Cheng, Zhenshun; Yu, Ting; Xia, Jiaan; Wei, Yuan; Wu, Wenjuan; Xie, Xuelei; Yin, Wen; Li, Hui; Liu, Min; Xiao, Yan; Gao, Hong; Guo, Li; Xie, Jungang; Wang, Guangfa; Jiang, Rongmeng; Gao, Zhancheng; Jin, Qi; Wang, Jianwei; Cao, Bin (24 January 2020). Clinical features of patients infected with 2019 novel coronavirus in Wuhan, China . The Lancet. 0 (10223): 497–506. doi:10.1016/S0140-6736(20)30183-5.

572) Hubei's coronavirus cases rise 10-fold after change in diagnostic criteria . South China Morning Post. 13 February 2020. Retrieved 13 February 2020.

573) Hui DS, I Azhar E, Madani TA, Ntoumi F, Kock R, Dar O, et al. (February 2020). The continuing 2019-nCoV epidemic threat of novel coronaviruses to global health - The latest 2019 novel

coronavirus outbreak in Wuhan, China . International Journal of Infectious Diseases. 91: 264–266. doi:10.1016/j.ijid.2020.01.009.

574) Hui DS, I Azhar E, Madani TA, Ntoumi F, Kock R, Dar O, et al. (February 2020). The continuing 2019-nCoV epidemic threat of novel coronaviruses to global health—The latest 2019 novel coronavirus outbreak in Wuhan, China . Int J Infect Dis. 91: 264–66. doi:10.1016/j.ijid.2020.01.009.

575) Hunter, Philip (2007). Inevitable or avoidable? Despite the lessons of history, the world is not yet ready to face the next great plague . EMBO Reports. 8 (6): 531–534. doi:10.1038/sj.embor.7400987.

576) Husband of Ontario's 5th COVID-19 case also tests positive for coronavirus . cbc.ca. 27 February 2020.

577) Huynh J, Li S, Yount B, Smith A, Sturges L, Olsen JC, et al. (December 2012). Evidence supporting a zoonotic origin of human coronavirus strain NL63 . Journal of Virology. 86 (23): 12816–25. doi:10.1128/JVI.00906-12.

578) Ian Sample (19 March 2020). Trials to begin on Covid-19 vaccine in UK next month . The Guardian.

579) Ibrahim, Arwa (19 February 2020). Two die of coronavirus in Iran, first fatalities in Middle East . Al Jazeera. Retrieved 19 February2020.

580) ICTV Taxonomy history: Orthocoronavirinae . International Committee on Taxonomy of Viruses (ICTV). Retrieved 2020-01-24.

581) In Memory of the 1910 Harbin Plague . Archived from the original on 2017-10-28. Retrieved 2011-01-14.

582) In view of Coronavirus outbreak, Air India suspends Delhi-Hong Kong flights from Feb 8 . mid-day. 4 February 2020. Archivedfrom the original on 7 February 2020. Retrieved 6 February 2020.

583) Inbjudan till pressträff: Ytterligare fall av covid-19 i flera regioner . Folkhalsomyndigheten. Retrieved 27 February 2020.

584) Inc, Gideon Informatics; Berger, Dr Stephen (2019). Measles: Global Status: 2019 edition. GIDEON Informatics Inc. p. 514. ISBN 978-1-4988-2461-3. Retrieved 12 February 2020.

585) India reports 2nd case of novel coronavirus: Ker govt says awaiting results from NIV . The Economic Times. 2 February 2020.

586) India struggles with deadly swine flu outbreak . BBC News. 20 February 2015. Archived from the original on 21 February 2015. Retrieved 21 February 2015.

587) India temporarily suspends e-visa for travellers from China - Times of India . The Times of India. Retrieved 5 February 2020.

588) Indonesia to stop flights to and from China amid coronavirus epidemic , The Jakarta Post, Jakarta, Indonesia, archived from the original on 2 February 2020, retrieved 2 February 2020

589) Infografía: así fue la ruta de llegada del coronavirus a México . Infobae. 28 February 2020. Retrieved 28 February 2020.

590) Inside Germany's coronavirus quarantine camp . Deutsche Welle. 3 February 2020. Archived from the original on 5 February 2020. Retrieved 3 February 2020.

591) International Committee on Taxonomy of Viruses (2010-08-24). ICTV Master Species List 2009—v10 (xls).

592) International Committee on Taxonomy of Viruses (ICTV). October 2018. Archived from the original on 2019-05-14. Retrieved 2020-01-24.

593) International Epidemiological Association (IEA)

594) International Health Regulations Emergency Committee on novel coronavirus in China . www.who.int.

595) Iran confirms 205 new coronavirus cases: State TV . Al Arabiya. Retrieved 29 February 2020.

596) Iran confirms 22 people dead, 141 infected with coronavirus: IRNA . Al Arabiya. Retrieved 27 February 2020.

597) Iran confirms 26 people dead, 245 infected with coronavirus: Fars . Al Arabiya. Retrieved 27 February 2020.

598) Iran COVID-19 Death Toll Reaches 8, With 43 Confirmed Infection Cases . Time. 23 February 2020. Retrieved 24 February2020.

599) Iran now says 5 dead out of 28 infected with new coronavirus . ABC News. 22 February 2020. Retrieved 22 February 2020.

600) Iran says coronavirus death toll reaches 16, 95 infected . National Post. 25 February 2020. Retrieved 26 February 2020.

601) Iran to Ban Exports of Face Masks . Financial Tribune. 4 February 2020. Retrieved 1 March 2020.

602) Iranian citizen tests positive for coronavirus in Estonia: Ifax . Al Arabiya. Retrieved 27 February 2020.

603) Iranian MP Mahmoud Sadeghi says he tested positive for coronavirus . Al Arabiya. Retrieved 25 February 2020.

604) Iranian official infected with coronavirus: Fars news agency . Al Arabiya. Retrieved 27 February 2020.

605) Iranian student, family of 4 are latest coronavirus cases in Iraq: Ministry . AL Arabiya. Retrieved 25 February 2020.

606) Iranian Vice President Masoumeh Ebtekar tests positive for coronavirus: Report . Al Arabiya. Retrieved 27 February 2020.

607) Iran's deputy health minister tests positive for coronavirus . Al Jazeera. Retrieved 25 February 2020.

608) Iraq announces sixth coronavirus case associated with travel to Iran . Al Arabiya. Retrieved 27 February 2020.

609) Israel confirms first coronavirus case as cruise ship returnee diagnosed . The Times of Israel. 21 February 2020. Archivedfrom the original on 21 February 2020. Retrieved 21 February 2020.

610) Israeli who recovered in Japan develops coronavirus again in Israel . Jerusalem Post. 28 February 2020. Retrieved 28 February2020.

611) Israeli who returned from Italy diagnosed with coronavirus: ministry . i24 news. Retrieved 27 February 2020.

612) Italian hospital saves Covid-19 patients lives by 3D printing valves for reanimation devices . 3D Printing Media Network. 14 March 2020. Retrieved 20 March 2020.

613) Italians scolded for flouting lockdown as death toll nears 3,000 . Pittsburgh Post-Gazette. Retrieved 20 March 2020.

614) Italy suspends all China flights after coronavirus cases confirmed in Rome . www.thelocal.it. Retrieved 31 January 2020.

615) Iwata-Yoshikawa N, Okamura T, Shimizu Y, Hasegawa H, Takeda M, Nagata N (March 2019). TMPRSS2 Contributes to Virus Spread and Immunopathology in the Airways of Murine Models after Coronavirus Infection . Journal of Virology. 93 (6). doi:10.1128/JVI.01815-18. PMC 6401451.

616) Jamaica imposes travel ban over coronavirus . jamaica-gleaner.com. 31 January 2020.
617) Jamaica turns back 19 Chinese nationals under coronavirus travel ban . www.loopjamaica.com.
618) Japan bars entry of foreign travelers from China's Hubei Province . Japan Today. Archived from the original on 1 February 2020. Retrieved 1 February 2020.
619) Japan confirms 70 new cases of COVID-19 from Diamond Princess cruise ship . The Japan Times Online. 16 February 2020. Retrieved 16 February 2020.
620) Japan confirms its first COVID-19 death: Health minister. CNA. 13 February 2020. Retrieved 13 February 2020.
621) Japan confirms six more cases from quarantined cruise ship as coronavirus deaths surpasses SARS' toll. The Japan Times. 9 February 2020. Retrieved 9 February 2020.
622) Japan reports second COVID-19 death, PM urges two-week curbs on sports events. CNA. 26 February 2020. Retrieved 26 February 2020.
623) Japan screening cruise ship for coronavirus. NHK World. Archived from the original on 4 February 2020. Retrieved 3 February 2020.
624) Japanese flu drug 'clearly effective' in treating coronavirus, says China . 18 March 2020.
625) Japanese woman confirmed as COVID-19 case for 2nd time, weeks after initial recovery . CNA. 27 February 2020. Retrieved 27 February 2020.
626) Jeong-ho, Lee; Zheng, William; Zhou, Laura (26 January 2020). Chinese scientists race to develop vaccine as coronavirus death toll jumps . South China Morning Post. Archived from the original on 26 January 2020. Retrieved 28 January 2020.
627) Ji Y, Ma Z, Peppelenbosch MP, Pan Q (February 2020). Potential association between COVID-19 mortality and health-care resource availability . Lancet Global Health. 8 (4): e480. doi:10.1016/S2214-109X(20)30068-1.

628) Jiang, Shibo; Lu, Lu; Du, Lanying (2013). Development of SARS vaccines and therapeutics is still needed . Future Virology. 8(1): 1–2. doi:10.2217/fvl.12.126.

629) jiemian.com (in Chinese). Archived from the original on 2 February 2020. Retrieved 2 February 2020.

630) Jin YH, Cai L, Cheng ZS, Cheng H, Deng T, Fan YP, et al. (February 2020). A rapid advice guideline for the diagnosis and treatment of 2019 novel coronavirus (2019-nCoV) infected pneumonia (standard version) . Military Medical Research. 7 (1): 4. doi:10.1186/s40779-020-0233-6.

631) Johansen, Bruce E. (2015). American Indian Culture: From Counting Coup to Wampum [2 volumes]: From Counting Coup to Wampum. ABC-CLIO. p. 88. ISBN 978-1-4408-2874-4.

632) John Snow (1855). On the mode of communication of cholera. John Churchill. Retrieved 29 March 2011.

633) Johnson, Carla K. (22 February 2020). US virus cases climb to 35 with return of cruise passengers . ABC News (US). Retrieved 22 February 2020.

634) Jones, Colin (1996). Plague and Its Metaphors in Early Modern France . Representations (53): 97–127. doi:10.2307/2928672. ISSN 0734-6018.

635) Jones, Jessica; Maler, Sandra (25 February 2020). Italian national becomes third case of coronavirus in Spain . Reuters. Retrieved 25 February 2020.

636) Joseph, Andrew (22 January 2020). WHO postpones decision on whether to declare China outbreak a global public health emergency . Stat News. Archived from the original on 22 January 2020. Retrieved 22 January 2020.

637) Coronavirus COVID-19 Global Cases by Johns Hopkins CSSE . ArcGIS. Johns Hopkins CSSE. Retrieved 2020-03-25.

638) Forgie S, Marrie TJ (February 2009). Healthcare-associated atypical pneumonia . Seminars in Respiratory and Critical Care Medicine. 30 (1): 67–85. doi:10.1055/s-0028-1119811.

639) Sexton NR, Smith EC, Blanc H, Vignuzzi M, Peersen OB, Denison MR (August 2016). Homology-Based Identification of a Mutation in the Coronavirus RNA-Dependent RNA Polymerase That

Confers Resistance to Multiple Mutagens . Journal of Virology. 90(16): 7415–28. doi:10.1128/JVI.00080-16. PMC 4984655. CoVs also have the largest known RNA virus genomes, ranging from 27 to 34 kb (31, 32), and increased fidelity in CoVs is likely required for the maintenance of these large genomes (14).

640) Junaidi, Ikram; Nagri, Jamil (27 January 2020). Coronavirus fear: GB seeks delay in opening of border crossing . Dawn. Archivedfrom the original on 27 January 2020. Retrieved 27 January 2020.

641) Jung SM, Akhmetzhanov AR, Hayashi K, Linton NM, Yang Y, Yuan B, et al. (February 2020). Real-Time Estimation of the Risk of Death from Novel Coronavirus (COVID-19) Infection: Inference Using Exported Cases . Journal of Clinical Medicine. 9 (2): 523. doi:10.3390/jcm9020523.

642) JUST NU: Bekräftat coronafall i Göteborg . GT. Retrieved 26 February 2020.

643) Kahn JS, McIntosh K (November 2005). History and recent advances in coronavirus discovery . The Pediatric Infectious Disease Journal. 24 (11 Suppl): S223-7, discussion S226. doi:10.1097/01.inf.0000188166.17324.60.

644) Kaiser Permanente launches first coronavirus vaccine trial . Kaiser Permanente, Washington Health Research Institute, Seattle. 16 March 2020. Retrieved 23 March 2020.

645) Kang L, Li Y, Hu S, Chen M, Yang C, Yang BX, et al. (March 2020). The mental health of medical workers in Wuhan, China dealing with the 2019 novel coronavirus . The Lancet. Psychiatry. 7(3): e14. doi:10.1016/S2215-0366(20)30047-X.

646) Kara Kostanich, KOMO News (30 January 2020). Washington activates emergency response center in wake of coronavirus . KOMO.

647) Karaalp, Haydar (24 February 2020). Iraq confirms first case of coronavirus . Anadolu Agency. Retrieved 24 February 2020.

648) Kaur, Karamjit; Goh, Timothy; Kurohi, Rei (12 February 2020). Coronavirus: 3 new cases in Singapore - 2 from Grace Assembly of God church, 1 from DBS . The Straits Times. Retrieved 12 February 2020.

649) Kaur, Karamjit; Koh, Fabian; Kurohi, Rei (14 February 2020). Coronavirus: 9 new cases confirmed in Singapore with 6 linked to Grace Assembly church . The Straits Times. Retrieved 14 February 2020.

650) Kazakhstan suspends transport links with China over virus . Reuters. Retrieved 30 January 2020.

651) Kelland K (2012-09-28). New virus not spreading easily between people: WHO . Reuters. Archived from the original on 2012-11-24. Retrieved 2013-03-16.

652) Kerala Confirms 3rd Case of Wuhan Coronavirus; India Issues Fresh Travel Advisory on China . News18. Archived from the original on 5 February 2020. Retrieved 8 February 2020.

653) Kertscher, Tom (23 January 2020). No, there is no vaccine for the Wuhan coronavirus . PolitiFact. Poynter Institute. Archived from the original on 7 February 2020. Retrieved 7 February 2020.

654) Khalik, Salma; Goh, Timothy (18 February 2020). Coronavirus: 4 new cases including 3 linked to Grace Assembly of God church; 21 cases now linked to the cluster . The Straits Times. Retrieved 18 February 2020.

655) Khalik, Salma; Goh, Timothy; Kurohi, Rei (30 January 2020). Wuhan virus: Every household in Singapore to get 4 masks; collection starts on Feb 1 . The Straits Times. Retrieved 30 January 2020.

656) Khan, E.; Siddiqui, J.; Shakoor, S.; Mehraj, V.; Jamil, B.; Hasan, R. (2007). Dengue outbreak in Karachi, Pakistan, 2006: Experience at a tertiary care center . Transactions of the Royal Society of Tropical Medicine and Hygiene. 101 (11): 1114–1119. doi:10.1016/j.trstmh.2007.06.016.

657) Kharpal, Arjun (5 February 2020). Tesla shares crater 17% after coronavirus delays Model 3 deliveries . CNBC. Archived from the original on 6 February 2020. Retrieved 5 February 2020.

658) Kim, Eun; Okada, Kaori; Kenniston, Tom; Raj, V. Stalin; Alhajri, Mohd M.; Farag, Elmoubasher A.B.A.; Alhajri, Farhoud; Osterhaus, Albert D.M.E.; Haagmans, Bart L.; Gambotto, Andrea (2014). Immunogenicity of an adenoviral-based Middle East Respiratory Syndrome coronavirus vaccine in BALB/C mice . Vaccine. 32 (45): 5975–5982. doi:10.1016/j.vaccine.2014.08.058.

659) Knesset to form special committee to deal with coronavirus . The Jerusalem Post JPost.com. Archived from the original on 6 February 2020. Retrieved 8 February 2020.
660) Ko WC, Rolain JM, Lee NY, Chen PL, Huang CT, Lee PI, Hsueh PR (March 2020). Arguments in favor of remdesivir for treating SARS-CoV-2 infections . International Journal of Antimicrobial Agents: 105933. doi:10.1016/j.ijantimicag.2020.105933.
661) Koh, B. K.; Ng, L. C.; Kita, Y.; Tang, C. S.; Ang, L. W.; Wong, K. Y.; James, L.; Goh, K. T. (2008). The 2005 dengue epidemic in Singapore: Epidemiology, prevention and control (PDF). Annals of the Academy of Medicine, Singapore. 37 (7): 538–45.
662) Kohn, George C. (2002). Encyclopedia of Plague and Pestilence: From Ancient Times to the Present. Princeton, New Jersey: Checkmark Books. p. 213. ISBN 978-0816048939.
663) Kollewe, Julia (18 March 2020). Experimental lung drug to be tested on UK coronavirus patients . The Guardian. ISSN 0261-3077. Retrieved 21 March 2020.
664) Korea Centers for Disease Control and Prevention. 23 March 2020. Retrieved 24 March 2020.
665) Korea to ban entry from China's Hubei province , Korea Herald, Seoul, South Korea, archived from the original on 6 February 2020, retrieved 2 February 2020
666) Koreas to temporarily close inter-Korean liaison office over virus concern . Korean Herald. 30 January 2020. Retrieved 30 January2020.
667) Koronawirus w Chinach. LOT zawiesza rejsy do Pekinu . Business Insider. 31 January 2020.
668) Kreative Lösungen gesucht . Startseite (in German).
669) Krebsbach, Suzanne (1996). The Great Charlestown Smallpox Epidemic of 1760 . The South Carolina Historical Magazine. 97 (1): 30–37. ISSN 0038-3082. JSTOR 27570134.
670) Kucharski, Adam J; Russell, Timothy W; Diamond, Charlie; Liu, Yang; Edmunds, John; Funk, Sebastian; Eggo, Rosalind M; Sun, Fiona; Jit, Mark; Munday, James D; Davies, Nicholas; Gimma, Amy; van Zandvoort, Kevin; Gibbs, Hamish; Hellewell, Joel; Jarvis,

Christopher I; Clifford, Sam; Quilty, Billy J; Bosse, Nikos I; Abbott, Sam; Klepac, Petra; Flasche, Stefan (March 2020). Early dynamics of transmission and control of COVID-19: a mathematical modelling study . The Lancet Infectious Diseases. doi:10.1016/S1473-3099(20)30144-4.

671) Kuhakan, Jiraporn (12 March 2020). 'Better than nothing': Thailand encourages cloth masks amid surgical mask shortage . Reuters. Thailand's health authorities are encouraging people to make cloth face masks at home to guard against the spread of the coronavirus amid a shortage of surgical masks. ... The droplet from coughing and sneezing is around five microns and we have tested already that cloth masks can protect against droplets bigger than one micron.

672) Kuhnke, Laverne. Lives at Risk: Public Health in Nineteenth-Century Egypt. ark.cdlib.org Archived 2008-11-20 at the Wayback Machine, Berkeley: University of California Press, c1990.

673) Kui L, Fang YY, Deng Y, Liu W, Wang MF, Ma JP, et al. (February 2020). Clinical characteristics of novel coronavirus cases in tertiary hospitals in Hubei Province . Chinese Medical Journal: 1. doi:10.1097/CM9.0000000000000744.

674) Kuo, Lily (10 February 2020). Xi Jinping appears in public as China returns to work after holiday . The Guardian. Retrieved 10 February 2020.

675) Kupferschmidt, Kai; CohenMar. 22, Jon; 2020; Pm, 3:28 (22 March 2020). WHO launches global megatrial of the four most promising coronavirus treatments . Science | AAAS. Retrieved 23 March 2020.

676) Kurohi, Rei (19 February 2020). Coronavirus: 3 new cases in Singapore, including one first warded as dengue patient; five more patients discharged . The Straits Times. Retrieved 19 February2020.

677) Kurohi, Rei (26 February 2020). Coronavirus: 2 new cases confirmed with no previous links, 4 more discharged . The Straits Times. Retrieved 26 February 2020.

678) Kurohi, Rei; Baharudin, Hariz (13 February 2020). Coronavirus: 8 new cases in Singapore, including 5 more linked to

Grace Assembly church, 1 linked to DBS case . The Straits Times. Retrieved 13 February 2020.

679) Kurohi, Rei; Goh, Timothy (17 February 2020). 1-year-old boy evacuated from Wuhan on Feb 9 among 2 new coronavirus cases in S'pore; the other case linked to DBS staff infected earlier . The Straits Times. Retrieved 17 February 2020.

680) Kuwait announces five coronavirus infections . english.alarabiya.net.

681) Kuwait confirms all 43 coronavirus cases recently traveled to Iran: Health Ministry . Al Arabiya. Retrieved 27 February 2020.

682) Laboratory testing for 2019 novel coronavirus (2019-nCoV) in suspected human cases . World Health Organization (WHO). Retrieved 13 March 2020.

683) Laboratory testing for 2019 novel coronavirus (2019-nCoV) in suspected human cases . Archived from the original on 21 February 2020. Retrieved 26 February 2020.

684) Laboratory testing of human suspected cases of novel coronavirus (nCoV) infection. Interim guidance, 10 January 2020 (PDF). Archived (PDF) from the original on 2020-01-20. Retrieved 2020-01-14.

685) Lai MM, Cavanagh D (1997). The molecular biology of coronaviruses . Advances in Virus Research. 48: 1–100. doi:10.1016/S0065-3527(08)60286-9. ISBN 9780120398485.

686) Lai, Chih-Cheng; Shih, Tzu-Ping; Ko, Wen-Chien; Tang, Hung-Jen; Hsueh, Po-Ren (1 March 2020). Severe acute respiratory syndrome coronavirus 2 (SARS-CoV-2) and coronavirus disease-2019 (COVID-19): The epidemic and the challenges . International Journal of Antimicrobial Agents. 55 (3): 105924. doi:10.1016/j.ijantimicag.2020.105924. ISSN 0924-8579.

687) Lai, Linette (27 January 2020). Wuhan virus: 2-3 students in every Singapore school likely to be affected by leave of absence . The Straits Times. Retrieved 16 February 2020.

688) Lai-Meng Looi; Kaw-Bing Chua (2007). Lessons from the Nipah virus outbreak in Malaysia (PDF). The Malaysian Journal of Pathology. Department of Pathology, University of Malaya and National Public Health Laboratory of the Ministry of Health,

Malaysia. 29 (2): 63–67. Archived (PDF) from the original on 30 August 2019.

689) Lào Cai tạm ngừng xuất, nhập cảnh khách du lịch qua cửa khẩu quốc tế . tuoitre.vn. Retrieved 27 January 2020.

690) Lapin, Tamar (5 February 2020). Wisconsin coronavirus case brings total infected in US to 12 . Archived from the original on 6 February 2020. Retrieved 8 February 2020.

691) Last flight from China arrives at Tijuana Airport as coronavirus causes evacuations . KUSI News. 27 January 2020. Retrieved 28 January 2020.

692) Latest: A new case positive for COVID19 in Malaysia . Ministry of Health (Malaysia) (retrieved from Twitter). 27 February 2020. Retrieved 27 February 2020.

693) Lau SK, Lee P, Tsang AK, Yip CC, Tse H, Lee RA, et al. (November 2011). Molecular epidemiology of human coronavirus OC43 reveals evolution of different genotypes over time and recent emergence of a novel genotype due to natural recombination . Journal of Virology. 85 (21): 11325–37. doi:10.1128/JVI.05512-11. PMC 3194943.

694) Lau SK, Li KS, Tsang AK, Lam CS, Ahmed S, Chen H, et al. (August 2013). Genetic characterization of Betacoronavirus lineage C viruses in bats reveals marked sequence divergence in the spike protein of pipistrellus bat coronavirus HKU5 in Japanese pipistrelle: implications for the origin of the novel Middle East respiratory syndrome coronavirus . Journal of Virology. 87 (15): 8638–50. doi:10.1128/JVI.01055-13.

695) Laude H, Rasschaert D, Delmas B, Godet M, Gelfi J, Charley B (June 1990). Molecular biology of transmissible gastroenteritis virus . Veterinary Microbiology. 23 (1–4): 147–54. doi:10.1016/0378-1135(90)90144-K.

696) Lauer, Stephen A.; Grantz, Kyra H.; Bi, Qifang; Jones, Forrest K.; Zheng, Qulu; Meredith, Hannah R.; Azman, Andrew S.; Reich, Nicholas G.; Lessler, Justin (10 March 2020). The Incubation Period of Coronavirus Disease 2019 (COVID-19) From Publicly Reported Confirmed Cases: Estimation and Application . Annals of Internal Medicine. doi:10.7326/M20-0504. ISSN 0003-4819.

697) Lebanon confirms three new cases of coronavirus, bringing total to seven: NNA . Reuters. 29 February 2020. Retrieved 29 February 2020.

698) Lee, Chermaine (15 February 2020). Malaysia confirms two more coronavirus cases . CNN. Archived from the original on 15 February 2020. Retrieved 15 February 2020.

699) Lee, Jaimy (7 March 2020). These nine companies are working on coronavirus treatments or vaccines — here's where things stand . MarketWatch. Retrieved 7 March 2020.

700) Lee, Jan (3 February 2020). K-pop group Winner's concert, Huayi events, Stage Club play cancelled or postponed due to coronavirus outbreak . The Straits Times. Retrieved 24 February 2020.

701) Lee, Se Young; Goh, Brenda (7 February 2020). Cruise line bans China citizens; Outrage as doctor who sounded coronavirus alarm dies . Reuters. Archived from the original on 6 February 2020. Retrieved 8 February 2020.

702) LeMay, Michael C. (2016). Global Pandemic Threats: A Reference Handbook: A Reference Handbook. ABC-CLIO. p. 227. ISBN 978-1-4408-4283-2. Retrieved 12 February 2020.

703) Leong, Trinna (6 February 2020). Malaysia confirms two new patients, bringing total coronavirus infections count to 14 . The Straits Times. Retrieved 6 February 2020.

704) Les autorités sanitaires de la Principauté ont été informées qu'une personne prise en charge dans la matinée et conduite au Centre Hospitalier Princesse Grace était positive au COVID 19. Son état de santé n'inspire pas d'inquiétude . Twitter. Gouvernement Monaco. 28 February 2020. Retrieved 28 February 2020.

705) Letko M, Marzi A, Munster V (2020). Functional assessment of cell entry and receptor usage for SARS-CoV-2 and other lineage B betacoronaviruses . Nature Microbiology: 1–8. doi:10.1038/s41564-020-0688-y.

706) Li G, De Clercq E (March 2020). Therapeutic options for the 2019 novel coronavirus (2019-nCoV) . Nature Reviews. Drug Discovery. 19 (3): 149–150. doi:10.1038/d41573-020-00016-0.

707) Li G, De Clercq E. Therapeutic options for the 2019 novel coronavirus (2019-nCoV). Nature Reviews Drug Discovery 2020 Feb doi:10.1038/d41573-020-00016-0

708) Li W, Farzan M, Harrison SC (September 2005). Structure of SARS coronavirus spike receptor-binding domain complexed with receptor . Science. 309 (5742): 1864–68. Bibcode:2005Sci...309.1864L. doi:10.1126/science.1116480.

709) Li XQ, Cai WF, Huang LF, Chen C, Liu YF, Zhang ZB, et al. (March 2020). [Comparison of epidemic characteristics between SARS in2003 and COVID-19 in 2020 in Guangzhou] . Zhonghua Liu Xing Bing Xue Za Zhi = Zhonghua Liuxingbingxue Zazhi (in Chinese). 41(5): 634–637. doi:10.3760/cma.j.cn112338-20200228-00209.

710) Li, G.; De Clercq, E. (2020). Therapeutic options for the 2019 novel coronavirus (2019-nCoV) . Nature Reviews. Drug Discovery. 19 (3): 149–150. doi:10.1038/d41573-020-00016-0.

711) Li, Qun; Guan, Xuhua; Wu, Peng; Wang, Xiaoye; Zhou, Lei; Tong, Yeqing; Ren, Ruiqi; Leung, Kathy S.M.; Lau, Eric H.Y.; Wong, Jessica Y.; Xing, Xuesen; Xiang, Nijuan; Wu, Yang; Li, Chao; Chen, Qi; Li, Dan; Liu, Tian; Zhao, Jing; Liu, Man; Tu, Wenxiao; Chen, Chuding; Jin, Lianmei; Yang, Rui; Wang, Qi; Zhou, Suhua; Wang, Rui; Liu, Hui; Luo, Yingbo; Liu, Yuan; Shao, Ge; Li, Huan; Tao, Zhongfa; Yang, Yang; Deng, Zhiqiang; Liu, Boxi; Ma, Zhitao; Zhang, Yanping; Shi, Guoqing; Lam, Tommy T.Y.; Wu, Joseph T.; Gao, George F.; Cowling, Benjamin J.; Yang, Bo; Leung, Gabriel M.; Feng, Zijian (29 January 2020). Early Transmission Dynamics in Wuhan, China, of Novel Coronavirus–Infected Pneumonia . New England Journal of Medicine. 0. doi:10.1056/NEJMoa2001316.

712) Li-Li; Wang, Ye-Ming; Wu, Zhi-Qiang; Xiang, Zi-Chun; Guo, Li; Xu, Teng; Jiang, Yong-Zhong; Xiong, Yan; Li, Yong-Jun; Li, Xing-Wang; Li, Hui; Fan, Guo-Hui; Gu, Xiao-Ying; Xiao, Yan; Gao, Hong; Xu, Jiu-Yang; Yang, Fan; Wang, Xin-Ming; Wu, Chao; Chen, Lan; Liu, Yi-Wei; Liu, Bo; Yang, Jian; Wang, Xiao-Rui; Dong, Jie; Li, Li; Huang, Chao-Lin; Zhao, Jian-Ping; Hu, Yi; Cheng, Zhen-Shun; Liu, Lin-Lin; Qian, Zhao-Hui; Qin, Chuan; Jin, Qi; Cao, Bin; Wang, Jian-Wei (11 February 2020). Identification of a novel coronavirus

causing severe pneumonia in human: a descriptive study. Chinese Medical Journal. doi:10.1097/CM9.0000000000000722. Limited data on coronavirus may be skewing assumptions about severity. STAT. 30 January 2020. Archived from the original on 1 February 2020. Retrieved 1 February 2020.

713) Lin, James T. Areddy and Liza (25 January 2020). WSJ News Exclusive | U.S. Plans to Evacuate Citizens From Epidemic-Stricken Chinese City . Wall Street Journal. ISSN 0099-9660. Retrieved 25 January 2020.

714) Lithuania confirms first coronavirus infection . Reuters. 28 February 2020. Retrieved 28 February 2020.

715) Liu P, Shi L, Zhang W, He J, Liu C, Zhao C, et al. (November 2017). Prevalence and genetic diversity analysis of human coronaviruses among cross-border children . Virology Journal. 14(1): 230. doi:10.1186/s12985-017-0896-0.

716) Liu R, Miller J (3 March 2020). China approves use of Roche drug in battle against coronavirus complications . Reuters. Retrieved 14 March 2020.

717) LIVE - CORONAVIRUS: SALGONO A 16 I CASI IN LIGURIA . Region Liguria. Retrieved 26 February 2020.

718) Lodge, T. (1603). A treatise of the plague: containing the nature, signes, and accidents of the same, with the certaine and absolute cure of the fevers, botches and carbuncles that raigne in these times. London: Edward White.

719) Los casos positivos de #coronavirus en España, a las 12 horas de hoy, 28 de febrero de 2020, son 32: (in Spanish) . Ministry of Health, Consumer Affairs and Social Welfare (Spain). 28 February 2020. Retrieved 28 February 2020.

720) Los tres portadores de coronavirus en México arribaron de Italia (in Spanish) . Expansión. 29 February 2020. Retrieved 29 February 2020.

721) Loss of sense of smell as marker of COVID-19 infection .

722) Low, Dominic (23 January 2020). Wuhan virus: Mindef issues two advisories for service personnel . The Straits Times. Retrieved 29 January 2020.

723) Lu H (28 January 2020). Drug treatment options for the 2019-new coronavirus (2019-nCoV) . Biosci Trends. 14 (1): 69–71. doi:10.5582/bst.2020.01020.
724) Lu X, Zhang L, Du H, Zhang J, Li YY, Qu J, et al. (18 March 2020). SARS-CoV-2 Infection in Children . New England Journal of Medicine. Massachusetts Medical Society. doi:10.1056/nejmc2005073. ISSN 0028-4793.
725) Lucey D, Sparrow A (14 January 2020). China Deserves Some Credit for Its Handling of the Wuhan Pneumonia . Foreign Policy. Retrieved 13 March 2020.
726) Luitwieler, Neal (30 January 2020). KLM staakt alsnog alle lijndiensten naar China . Luchtvaartnieuws.
727) Lum, Alvin; Cheung, Elizabeth (6 February 2020). New confirmed coronavirus case in Hong Kong, as two more test positive . South China Morning Post. Archived from the original on 6 February 2020. Retrieved 6 February 2020.
728) Lum, Alvin; Low, Zoe (24 February 2020). Coronavirus: two more Hong Kong evacuees from Diamond Princess cruise ship confirmed with infection, bringing city's total to 81 . South China Morning Post. Retrieved 24 February 2020.
729) Lum, Alvin; Ting, Victor (17 February 2020). Coronavirus: Hospital Authority reveals staff were at a meeting attended by two engineers who later tested positive, as Hong Kong confirms 60th case . South China Morning Post. Retrieved 17 February 2020.
730) Luna, Franco (2 February 2020). DOH confirms second case of novel coronavirus in Philippines . The Philippine Star. Archivedfrom the original on 3 February 2020. Retrieved 2 February 2020.
731) Lynch, Lily (5 December 2015). Odessa, 1812: Plague and Tyranny at the Edge of the Empire . Balkanist. Archived from the original on 15 March 2020.
732) Macau to deny entry to visitors from Hubei unless they can show they are free of Wuhan virus . Channel News Asia. Retrieved 27 January 2020.

733) Madagascar: eighteen dead from Bubonic Plague, five in hospital since 1 January 2008 . Archived from the original on May 9, 2009.

734) Maddicott, J. R. (1 August 1997). Plague in seventh century England . Past & Present. 156 (1): 7–54. doi:10.1093/past/156.1.7. ISSN 0031-2746.

735) Maham Abedi (23 March 2020). Canada to spend $192M on developing COVID-19 vaccine . Global News. Retrieved 24 March 2020.

736) Mahtani S, Berger M, O'Grady S, Iati M (6 February 2020). Hundreds of evacuees to be held on bases in California; Hong Kong and Taiwan restrict travel from mainland China . The Washington Post. Archived from the original on 7 February 2020. Retrieved 11 February 2020.

737) Majka zaražene braće radi u ministarstvu: 'Nije bila na poslu i nema dodatnih mjera' . VL. Retrieved 26 February 2020.

738) Makeshift hospital delivered in northwest China city amid coronavirus battle - Xinhua | English.news.cn . www.xinhuanet.com. Retrieved 17 February 2020.

739) Malaria drug touted as coronavirus treatment by Trump and Elon Musk can be deadly, China finds . Fortune. Retrieved 23 March2020.

740) Malaria Epidemic Sweeps Northeast India . Archived from the original on 2017-02-03. Retrieved 2008-07-01.

741) Malaysia confirms 2 more cases of COVID-19 . CNA. 15 February 2020. Retrieved 15 February 2020.

742) Maldives bans travellers from China over virus fears, minister says , Reuters, archived from the original on 3 February 2020, retrieved 3 February 2020

743) Mali: Yellow fever epidemic in Kayes . 2005-11-11. Archivedfrom the original on 2009-01-12. Retrieved 2008-11-24.

744) Man infected with coronavirus in Germany after Italy trip . The Jerusalem Post. 26 February 2020. Retrieved 26 February 2020.

745) Manancourt, Vincent (10 March 2020). Coronavirus tests Europe's resolve on privacy . POLITICO. Retrieved 20 March2020.

746) Mann, Charles C. (December 2005). Native intelligence . Archived from the original on 2018-11-23. Retrieved 2018-10-17.

747) Maragakis LL. Coronavirus, Social Distancing and Self Quarantine . www.hopkinsmedicine.org. John Hopkins University.

748) Maresca, Thomas (25 February 2020). South Korea coronavirus cases near 1,000; 11 dead . UPI. Retrieved 26 February 2020.

749) Mark Thorson (29 January 2020). WHO praises China's response to coronavirus, will reconvene expert panel .

750) Marr, John S.; Cathey, John T. (2010). New Hypothesis for Cause of Epidemic among Native Americans, New England, 1616–1619 . Emerging Infectious Diseases. 16 (2): 281–286. doi:10.3201/eid1602.090276.

751) Martin PM, Martin-Granel E (June 2006). 2,500-year evolution of the term epidemic (PDF). Emerging Infect. Dis. 12 (6): 976–80. doi:10.3201/eid1206.051263.

752) Martin TR, Pistorese BP, Chi EY, Goodman RB, Matthay MA (November 1989). Effects of leukotriene B4 in the human lung. Recruitment of neutrophils into the alveolar spaces without a change in protein permeability . J. Clin. Invest. 84 (5): 1609–19. doi:10.1172/JCI114338. PMC 304027.

753) Masters PS (2006-01-01). The molecular biology of coronaviruses . Advances in Virus Research. Academic Press. 66: 193–292. doi:10.1016/S0065-3527(06)66005-3. ISBN 9780120398690.

754) Maugh, Thomas. An Empire's Epidemic . www.ph.ucla.edu. Retrieved 20 March 2020.

755) Mazan, Ryan; Gagnon, Alain; Desjardins, Bertrand (2009). The Measles Epidemic of 1714–1715 in New France . Canadian Studies in Population. 36 (3–4): 295–323. doi:10.25336/P63P5Q.

756) Mazumdar, Tulip (30 January 2020). Coronavirus: Scientists race to develop a vaccine . BBC News. Archived from the original on 30 January 2020. Retrieved 3 February 2020.

757) McCredie, J. (2009). Dengue fever epidemic hits northern Australia . BMJ. 338: b967. doi:10.1136/bmj.b967.

758) McCurry, Justin; Ratcliffe, Rebecca (17 February 2020). Diamond Princess cruise ship in Japan confirms 99 new coronavirus cases . The Guardian. Retrieved 17 February 2020.

759) McDonald, Jessica (24 January 2020). Social Media Posts Spread Bogus Coronavirus Conspiracy Theory . FactCheck.org. Annenberg Public Policy Center. Archived from the original on 6 February 2020. Retrieved 8 February 2020.

760) McNeil, Sam; Wang, Penny Yi; Kurtenbach, Elaine (26 January 2020). China Virus Outbreak Revives Calls to Stop Wildlife Trade . U.S.News. Archived from the original on 26 January 2020. Retrieved 26 January 2020.

761) Medicago announces production of a viable vaccine candidate for COVID-19 . Business Wire. 12 March 2020. Retrieved 24 March 2020.

762) Megadrought and Megadeath in 16th Century Mexico. National Institute of Health. April 8, 2002.

763) Mehta P, McAuley DF, Brown M, et al. (16 March 2020). COVID-19: consider cytokine storm syndromes and immunosuppression . The Lancet. doi:10.1016/S0140-6736(20)30628-0.

764) Merck Veterinary Manual . Merck Veterinary Manual. Archived from the original on 2019-12-13. Retrieved 2020-01-24.

765) Mexico confirms first 2 cases of new coronavirus . Los Angeles Times. 28 February 2020. Retrieved 28 February 2020.

766) mhlw.go.jp (in Japanese). Ministry of Health, Labour and Welfare. Archived from the original on 1 February 2020. Retrieved 1 February 2020.

767) Microbe hunters, then and now. Medi-Ed Press. 1996. p. 23. ISBN 978-0-936741-11-6.

768) Middle East Respiratory Syndrome coronavirus (MERS-CoV) – Saudi Arabia. World Health Organisation (2015-06-11). Retrieved on 2015-06-20 . Archived from the original on 2015-06-16. Retrieved 2015-06-20.

769) Middle East respiratory syndrome coronavirus (MERS-CoV) – United Arab Emirates . World Health Organization. 31 January 2020.

770) Middle East respiratory syndrome coronavirus (MERS-CoV) . WHO. Archived from the original on 2019-10-18. Retrieved 2019-12-10.
771) Middle East respiratory syndrome coronavirus (MERS-CoV)—Republic of Korea . World Health Organization. Retrieved 2016-12-01.
772) Mild/Moderate 2019-nCoV Remdesivir RCT - Full Text View - ClinicalTrials.gov . clinicaltrials.gov. Retrieved 20 March 2020.
773) Ministère de la Santé Publique et de la Population official cholera report 22 March 2011, www.mspp.gouv.ht Archived 2016-12-28 at the Wayback Machine.
774) Ministério da Saúde investiga possível paciente com coronavírus em SP; caso foi para contraprova . G1. 25 February 2020.
775) Ministry of Health and Welfare (South Korea) (in Korean). 26 February 2020. Retrieved 26 February 2020.
776) Ministry of Health registered first two Novel #Coronavirus (COVID-2019) cases for Omani women coming from #Iran . Ministry of Health (Oman) (retrieved from Twitter). 24 February 2020. Retrieved 24 February 2020.
777) Miri, Seyyed Mohammad; Roozbeh, Fatemeh; Omrani Rad, Ali; Alavian, Seyed Moayed (16 March 2020). Panic of Buying Toilet Papers: A Historical Memory or a Horrible Truth? Systematic Review of Gastrointestinal Manifestations of COVID-19 . Hepatitis Monthly. In Press (In Press). doi:10.5812/hepatmon.102729. ISSN 1735-143X.
778) Mishra, Manas (29 January 2020). Orr, Bernard; Kuber, Shailesh (eds.). Johnson & Johnson working on vaccine for deadly coronavirus . Reuters. Archived from the original on 29 January 2020. Retrieved 19 February 2020.
779) Modjarrad, Kayvon; Roberts, Christine C.; Mills, Kristin T.; Castellano, Amy R.; Paolino, Kristopher; Muthumani, Kar; Reuschel, Emma L.; Robb, Merlin L.; Racine, Trina; Oh, Myoung-don; Lamarre, Claude; Zaidi, Faraz I.; Boyer, Jean; Kudchodkar, Sagar B.; Jeong, Moonsup; Darden, Janice M.; Park, Young K.; Scott, Paul T.; Remigio, Celine; Parikh, Ajay P.; Wise, Megan C.; Patel, Ami; Duperret, Elizabeth K.; Kim, Kevin Y.; Choi, Hyeree; White, Scott; Bagarazzi, Mark; May, Jeanine M.; Kane, Deborah; et al. (2019).

Safety and immunogenicity of an anti-Middle East respiratory syndrome coronavirus DNA vaccine: a phase 1, open-label, single-arm, dose-escalation trial . The Lancet Infectious Diseases. 19 (9): 1013–1022. doi:10.1016/S1473-3099(19)30266-X.

780) Monash Virtual Laboratory - Simulations of epidemic spread across a landscape

781) Mongolia shuts universities, border crossings to halt virus spread . finance.yahoo.com. Retrieved 27 January 2020.

782) Morens, David M. (2015). The Past Is Never Dead—Measles Epidemic, Boston, Massachusetts, 1713 – Volume 21, Number 7 – July 2015 – Emerging Infectious Diseases journal – CDC . Emerging Infectious Diseases Journal. 21 (7): 1257–60. doi:10.3201/eid2107.150397.

783) Moscow deploys facial recognition technology for coronavirus quarantine . Reuters. 21 February 2020. Retrieved 20 March2020.

784) Một Việt kiều ở TP.HCM dương tính virus corona sau khi quá cảnh Vũ Hán . Tuổi Trẻ (in Vietnamese). Ho Chi Minh City, Vietnam. 2 February 2020. Archived from the original on 2 February 2020. Retrieved 2 February 2020.

785) Mourners die as fever grips Congo. Archived 2012-07-30 at WebCite Sydney Morning Herald, August 30, 2007

786) Mui, Rachel (29 January 2020). Dasin Retail Trust shortens hours, shuts businesses at China malls . The Business Times. Retrieved 16 February 2020.

787) Multicenter collaboration group of Department of Science Technology of Guangdong Province Health Commission of Guangdong Province for chloroquine in the treatment of novel coronavirus pneumonia (February 2020). [Expert consensus on chloroquine phosphate for the treatment of novel coronavirus pneumonia] . Zhonghua Jie He He Hu Xi Za Zhi = Zhonghua Jiehe He Huxi Zazhi = Chinese Journal of Tuberculosis and Respiratory Diseases. 43: E019. doi:10.3760/cma.j.issn.1001-0939.2020.0019.

788) Murphy FA, Gibbs EP, Horzinek MC, Studdart MJ (1999). Veterinary Virology. Boston: Academic Press. pp. 495–508. ISBN 978-0-12-511340-3.

789) Murphy, Simon; Sabbagh, Dan (30 January 2020). Coronavirus: Britons to be quarantined in Wirral after arriving from Wuhan . The Guardian – via www.theguardian.com.
790) Murray J (2014-04-16). What's New With Ferret FIP-like Disease? (xls). Archived from the original on 2014-04-24. Retrieved 2014-04-24.
791) Murthy S, Gomersall CD, Fowler RA (11 March 2020). Care for Critically Ill Patients With COVID-19. JAMA. doi:10.1001/jama.2020.3633.
792) My Hand-Washing Song: Readers Offer Lyrics For A 20-Second Scrub . NPR.org. Retrieved 20 March 2020.
793) Mzezewa, Tariro; Arnot, Mike; Chokshi, Niraj (31 January 2020). Delta and American Suspend Flights to China . The New York Times.
794) N. Korea quarantines suspected coronavirus cases in Sinuiju . Daily NK. 28 January 2020. Archived from the original on 30 January 2020. Retrieved 30 January 2020.
795) NA (1849). On the Influenza, or Epidemic Catarrhal Fever of 1847–8 . The American Journal of the Medical Sciences. 18(35): 148–154. doi:10.1097/00000441-184907000-00018.
796) Naming the coronavirus disease (COVID-19) and the virus that causes it . World Health Organization (WHO). Archived from the original on 28 February 2020. Retrieved 28 February 2020.
797) Naming the coronavirus disease (COVID-19) and the virus that causes it . World Health Organization (WHO). Retrieved 13 March 2020.
798) National Health Commission of the People's Republic of China. Archived from the original on 26 January 2020. Retrieved 9 February 2020.
799) National Health Commission of the People's Republic of China. Archived from the original on 26 January 2020. Retrieved 9 February 2020.
800) National Health Commission of the People's Republic of China. Retrieved 9 February 2020.

801) NBC15 Staff. Dane County resident is Wisconsin's first confirmed coronavirus case . www.nbc15.com. Archived from the original on 6 February 2020. Retrieved 8 February 2020.

802) nCoV-2019: preliminary estimates of the confirmed-case-fatality-ratio and infection-fatality-ratio, and initial pandemic risk assessment . institutefordiseasemodeling.github.io. Retrieved 1 March 2020.

803) NDTV Report . Archived from the original on 2009-02-25. Retrieved 2009-02-23.

804) Nebehay S, Kelland K, Liu R (5 February 2020). WHO: 'no known effective' treatments for new coronavirus . Thomson Reuters. Archived from the original on 5 February 2020. Retrieved 5 February 2020.

805) Nepal-China Rasuwagadhi Border Shut For 15 Days Amid Coronavirus Threat . NDTV Media. Retrieved 29 January 2020.

806) Neuman BW, Adair BD, Yoshioka C, Quispe JD, Orca G, Kuhn P, et al. (August 2006). Supramolecular architecture of severe acute respiratory syndrome coronavirus revealed by electron cryomicroscopy . Journal of Virology. 80 (16): 7918–28. doi:10.1128/JVI.00645-06. PMC 1563832. Particle diameters ranged from 50 to 150 nm, excluding the spikes, with mean particle diameters of 82 to 94 nm.

807) Neuman BW, Kiss G, Kunding AH, Bhella D, Baksh MF, Connelly S, et al. (April 2011). A structural analysis of M protein in coronavirus assembly and morphology . Journal of Structural Biology. 174 (1): 11–22. doi:10.1016/j.jsb.2010.11.021.

808) New coronavirus . Federal Office of Public Health (Switzerland). 28 February 2020. Retrieved 28 February 2020.

809) New coronavirus cases in Kuwait associated with Iran travel bring total to 25 . Al Arabiya. 26 February 2020. Retrieved 26 February2020.

810) New coronavirus cases on Diamond Princess . NHK. 15 February 2020. Retrieved 16 February 2020.

811) New coronavirus stable for hours on surfaces . National Institutes of Health. Retrieved 23 March 2020.

812) New COVID-19 cases reported across Japan as 88 more aboard Diamond Princess test positive . The Japan Times. 18 February 2020. Retrieved 18 February 2020.

813) New infection cases reported in Japan Saturday . NHK World. 22 February 2020. Retrieved 24 February 2020.

814) New SARS-like virus found in Middle East . Al-Jazeera. 2012-09-24. Archived from the original on 2013-03-09. Retrieved 2013-03-16.

815) New virus cases in UK, Germany, Italy put Europe total at 31 . AP NEWS. 6 February 2020. Archived from the original on 6 February 2020. Retrieved 8 February 2020.

816) New Zealand confirms case of Covid-19 coronavirus . Radio New Zealand. 28 February 2020. Retrieved 28 February 2020.

817) New, Deadly Flu Strain Detected in Albany Co . Schenectady Gazette. Associated Press. January 24, 1975. p. 3. Retrieved August 3, 2012.

818) Newburger, Emma (29 February 2020). Washington state confirms first US death from coronavirus . CNBC. Retrieved 29 February 2020.

819) Newey, Sarah; Kelly-Linden, Jordan (21 February 2020). Coronavirus latest news: South Korea in lockdown after soar in new cases . The Telegraph. Retrieved 21 February 2020.

820) Newman, Kira L. S. (2012). Shutt up: bubonic plague and quarantine in early modern England . Journal of Social History. 45(3): 809–834. doi:10.1093/jsh/shr114. ISSN 0022-4529.

821) News, A. B. C. Coronavirus death toll hits 812, surpassing SARS fatalities . ABC News.

822) News, A. B. C. Northern Italy closes schools and suspends gatherings as coronavirus cases mount . ABC News.

823) News, Lufthansa (29 January 2020). Lufthansa, SWISS und Austrian Airlines werden ihre Flüge von und nach China (Festland) bis 9. Februar aussetzen. Aus operativen Gründen ist die Buchungsannahme für Flüge nach China (Festland) bis Ende Februar gestoppt. Hongkong wird weiter wie geplant angeflogen.pic.twitter.com/CxfHuuTJ7s . @lufthansaNews (in German). Retrieved 29 January 2020.

824) News, Taiwan. Taiwan announces 3 more cases of Wuhan virus,... Taiwan News.

825) News, Taiwan. Taiwan records 18th coronavirus case, patient... Taiwan News.

826) Ng, Kang-chung; Cheung, Tony; Cheung, Elizabeth (12 February 2020). Coronavirus: Hong Kong discharges first recovered patient as 10 others from city confirmed to be infected in Hubei . South China Morning Post. Retrieved 13 February 2020.

827) Ng, Kang-chung; Wong, Brian; Low, Zoe (18 February 2020). Coronavirus: Hong Kong confirms 62nd infection, while mother-in-law of engineer in earlier case tests positive . South China Morning Post. Retrieved 18 February 2020.

828) Người dân lên Móng Cái để sang Trung Quốc được yêu cầu quay về (in Vietnamese). 31 January 2020. Archived from the original on 1 February 2020. Retrieved 31 January 2020.

829) Nguyen, Ngoc TB; Pham, Hau V.; Hoang, Cuong Q.; Nguyen, Tien M.; Nguyen, Long T.; Phan, Hung C.; Phan, Lan T.; Vu, Long N.; Tran Minh, Nguyen N. (2014). Epidemiological and clinical characteristics of children who died from hand, foot and mouth disease in Vietnam, 2011 . BMC Infectious Diseases. 14: 341. doi:10.1186/1471-2334-14-341.

830) Nienaber, Michael; Reese, Chris (12 February 2020). Germany confirms two more cases of coronavirus, raising overall number to 16 . Reuters. Retrieved 12 February 2020.

831) Nigeria cholera outbreak kills 400 . 2001-11-26. Archivedfrom the original on 19 December 2003. Retrieved 16 November2014.

832) Nigeria confirms first coronavirus case in Sub-Saharan Africa . France24. 28 February 2020. Retrieved 28 February 2020.

833) NIH clinical trial of investigational vaccine for COVID-19 begins . US National Institutes of Health. 16 March 2020. Retrieved 17 March 2020.

834) Nikel, David (29 February 2020). Coronavirus in Norway: The Latest News on the COVID-19 Outbreak . Life in Norway. Retrieved 29 February 2020.

835) Nine in family confirmed with coronavirus after Hong Kong hotpot dinner . South China Morning Post. 9 February 2020.

836) Ninth case of fast-moving coronavirus confirmed in U.S. Reuters. 2 February 2020. Archived from the original on 3 February 2020. Retrieved 3 February 2020.
837) Nipah virus contained, last two positive cases have recovered: Kerala Health Min . The News Minute. 2018-06-11. Retrieved 2020-02-16.
838) No Known Effective Therapeutics Against Coronavirus : WHO . NDTV.com. Archived from the original on 6 February 2020. Retrieved 6 February 2020.
839) Northern border province suspends tourists' entry, exit through int'l border gate . www.nytimes.com. Retrieved 27 January 2020.
840) Northern border province suspends tourists' entry, exit through int'l border gate . en.vietnamplus.vn. Retrieved 27 January 2020.
841) Northern Ireland diagnoses first coronavirus case . BBC News. 27 February 2020. Retrieved 27 February 2020.
842) Northwell Health Initiates Clinical Trials of 2 COVID-19 Drugs . 21 March 2020.
843) Northwell Health Initiates Clinical Trials of 2 COVID-19 Drugs . 21 March 2020.
844) Norway detects its first case of coronavirus . Reuters. Retrieved 26 February 2020.
845) Nouveau coronavirus—Point de situation : Un nouveau cas d'infection confirmé Archived 8 June 2013 at the Wayback Machine (Novel coronavirus—Status report: A new case of confirmed infection) 12 May 2013, social-sante.gouv.fr
846) Novel Coronavirus (2019-nCoV) . Centers for Disease Control and Prevention. Archivedfrom the original on 14 February 2020. Retrieved 15 February 2020.
847) Novel Coronavirus (2019-nCoV) . Centers for Disease Control and Prevention. 11 February 2020. Archived from the original on 7 March 2020. Retrieved 18 February 2020. The virus is thought to spread mainly from person-to-person ... through respiratory droplets produced when an infected person coughs or sneezes.
848) Novel Coronavirus (2019-nCoV) Situation Summary . Centers for Disease Control and Prevention. 30 January 2020. Archived from the original on 26 January 2020. Retrieved 30 January 2020.

849) Novel Coronavirus (COVID-19) Situation Report – 24 (PDF). WHO. 13 February 2020.
850) Novel Coronavirus (COVID-19) Situation Report – 25 (PDF). WHO. 14 February 2020.
851) Novel Coronavirus (COVID-19) Situation Report – 26 (PDF). WHO. 15 February 2020.
852) Novel Coronavirus (COVID-19) Situation Report – 27 (PDF). WHO. 16 February 2020.
853) Novel Coronavirus 2019, Wuhan, China | CDC . www.cdc.gov. 2020-01-23. Archived from the original on 2020-01-20. Retrieved 2020-01-23.
854) Novel coronavirus claims first foreign victims as US citizen, Japanese man die in Wuhan . CNA. Retrieved 8 February 2020.
855) Novel Coronavirus in China - Warning - Level 3, Avoid Nonessential Travel - Travel Health Notices | Travelers' Health | CDC . wwwnc.cdc.gov.
856) Novel Coronavirus in Hubei Province, China - Warning - Level 3, Avoid Nonessential Travel - Travel Health Notices | Travelers' Health | CDC . wwwnc.cdc.gov. Retrieved 27 January 2020.
857) Novel Coronavirus infection, 2019 (Wuhan, China): Outbreak update . Canada.ca. 2020-01-21.
858) Novel coronavirus infection—update . World Health Association. 2013-05-22. Archived from the original on 2013-06-07. Retrieved 2013-05-23.
859) Novel coronavirus named 'Covid-19': WHO . TODAYonline. Retrieved 11 February 2020.
860) Novel Coronavirus Pneumonia Diagnosis and Treatment Plan (Provisional 7th Edition) . China Law Translate. 4 March 2020.
861) Novel Coronavirus Pneumonia Diagnosis and Treatment Plan (Provisional 7th Edition) . China Law Translate. 4 March 2020.
862) Novel coronavirus to be called COVID-19, says WHO . CNA. 11 February 2020. Retrieved 11 February 2020.
863) Novel Coronavirus vaccine manufacturing contract signed . The Jenner Institute Laboratories, University of Oxford, England. 7 February 2020.

864) Novel Coronavirus(2019-nCoV) Situation Report - 12 (PDF), archived (PDF) from the original on 6 February 2020, retrieved 8 February 2020
865) Novel Coronavirus(2019-nCoV) Situation Report - 14 (PDF), archived (PDF) from the original on 6 February 2020, retrieved 8 February 2020
866) Novel Coronavirus(2019-nCoV) Situation Report - 15 (PDF), archived (PDF) from the original on 7 February 2020, retrieved 8 February 2020
867) Novel Coronavirus(2019-nCoV) Situation Report - 17 (PDF)
868) Novel Coronavirus(2019-nCoV) Situation Report - 19 (PDF)
869) Novel Coronavirus(2019-nCoV) Situation Report - 20 (PDF)
870) Novel Coronavirus(2019-nCoV) Situation Report - 21 (PDF)
871) Novel Coronavirus(2019-nCoV) Situation Report – 22 (PDF). World Health Organization. 11 February 2020. Retrieved 11 February 2020.
872) Novel Coronavirus(2019-nCoV) Situation Report—10 (PDF). World Health Organization (WHO). 30 January 2020. Retrieved 15 March 2020.
873) Novel Coronavirus(2019-nCoV) Situation Report-13 (PDF). Archived (PDF) from the original on 6 February 2020. Retrieved 8 February 2020.
874) Novel Coronavirus(2019-nCoV)Situation Report –18-ERRAT (PDF). WHO. 7 February 2020.
875) Novel Coronavirus(2019-nCoV)Situation Report –23-ERRAT (PDF). WHO. 12 February 2020.
876) Novel coronavirus: 2 new confirmed cases in Singapore, both with no recent travel history to China . TODAYonline. 11 February 2020. Retrieved 11 February 2020.
877) Novel coronavirus: 3 new cases confirmed, all have not recently travelled to China . TODAYonline. 9 February 2020. Retrieved 9 February 2020.
878) Novel Coronavirus—COVID-19: What Emergency Clinicians Need to Know . www.ebmedicine.net. Retrieved 9 March 2020.

879) Novi zaraženi od koronavirusa bili su u bliskom kontaktu s prvim i trećim oboljelim . VL. 28 February 2020. Retrieved 28 February2020.

880) NRW meldet 14 neue Infektionen . ntv. Retrieved 27 February2020.

881) Number of coronavirus cases in Italy jumps to 79 . TeleTrader. 22 February 2020. Retrieved 23 February 2020.

882) Number Of Dead From Virus Surges As China Changes Way It Counts Cases . NDTV.com. Retrieved 13 February 2020.

883) Number of Kuwait coronavirus cases rises to eight: KUNA . Al Arabiya. Retrieved 25 February 2020.

884) Número de casos informados de artritis epidémica chikungunya en las Américas – SE 5 (February 6, 2015) . Pan American Health Organization. Archived from the original on February 18, 2015. Retrieved February 11, 2015.

885) Nussey, Sam; Maxwell, Kenneth (8 February 2020). Japan confirms 3 more coronavirus cases on cruise liner; total now 64 . Reuters. Retrieved 8 February 2020.

886) NY COVID-19 cases surge; Javits Center to house temporary hospitals . 23 March 2020.

887) NZ to close doors on foreign travellers from China . Radio New Zealand. 2 February 2020. Archived from the original on 2 February 2020.

888) O'Carroll, Chad. North Korea suspends Vladivostok flights, expands virus containment measures (2020-02-01). NKNEWS. Archived from the original on 2 February 2020. Retrieved 2 February 2020.

889) Odigwe, C. (2009). West Africa has worst meningitis epidemic for 10 years . BMJ. 338: b1638. doi:10.1136/bmj.b1638.

890) Odisha grapples with jaundice outbreak . Deccan Herald. 17 February 2015. Archived from the original on 17 February 2015. Retrieved 17 February 2015.

891) Oficial: Primul român confirmat cu coronavirus . Antena3.Ro. Retrieved 26 February 2020.

892) Oficial: Primul român confirmat cu coronavirus . Reuters. Retrieved 26 February 2020.

893) Oguro, Junko (15 February 2020). Japan has nine more confirmed coronavirus cases . CNN. Retrieved 15 February 2020.
894) Olson, PE; Hames, CS; Benenson, AS; Genovese, EN (1996). The Thucydides syndrome: Ebola déjà vu? (or Ebola reemergent?) . Emerging Infect. Dis. 2 (2): 155–156. doi:10.3201/eid0202.960220. PMC 2639821.
895) Oltermann, Philip (15 March 2020). Trump 'offers large sums' for exclusive access to coronavirus vaccine . The Guardian.
896) Oman reports two more fresh Coronavirus cases . Oman Daily Observer. 25 February 2020. Retrieved 25 February 2020.
897) One more coronavirus case reported in Oman . Times of Oman. 27 February 2020. Retrieved 27 February 2020.
898) One of Korean Air's cabin crew tests positive for coronavirus . The Straits Times. 25 February 2020. Retrieved 29 February 2020.
899) Ontario confirms eighth coronavirus case is Toronto man who travelled to Egypt National Post . 29 February 2020.
900) Ontario confirms new case of coronavirus, patient had travelled to Iran . National Post. 26 February 2020. Retrieved 26 February2020.
901) Ontario Confirms New Positive Case of COVID-19 . ontario.ca(Press release). Queen's Printer for Ontario. 28 February 2020.
902) Operations Dashboard for ArcGIS . who.maps.arcgis.com. Archived from the original on 21 August 2019. Retrieved 20 February 2020.
903) Osumi, Magdalena (12 February 2020). Japan faces COVID-19 test kit shortage as infections on ship rise . The Japan Times Online. Retrieved 13 February 2020.
904) Our Heritage Saved: St Roque Chapel. The Malta Independent. 30 May 2007. Archived from the original on 13 March 2020.
905) Pacheco, Daniela Alexandra de Meneses Rocha; Rodrigues, Acácio Agostinho Gonçalves; Silva, Carmen Maria Lisboa da (October 2016). Ebola virus – from neglected threat to global emergency state . Revista da Associação Médica Brasileira. 62 (5): 458–467. doi:10.1590/1806-9282.62.05.458.

906) Pacific countries, already hard hit by epidemics, take extreme coronavirus measures . Washington Post. 29 January 2020. Retrieved 29 January 2020.

907) Paksoy, Yunus. German telecom giant shares private data with government amid privacy fears . trtworld. Retrieved 20 March2020.

908) PAL, Cebu Pacific suspend flights between Philippines, China . ABS-CBN News. 2 February 2020. Archived from the original on 2 February 2020. Retrieved 2 February 2020.

909) Panamá aumenta alerta sanitaria por coronavirus . TVN (in Spanish). 21 January 2020. Archived from the original on 22 January 2020. Retrieved 22 January 2020.

910) Panama Canal says ships must report contact with coronavirus countries 31 January 2020 www.reuters.com accessed 18 March 2020

911) Panamá habilita sala para atención de casos sospechosos de coronavirus 29 January 2020 www.telemetro.com accessed 18 March 2020

912) Pandemic Epidemic Diseases news: Infectious disease outbreaks reported in the Eastern Mediterranean region in 2018 Archived29 January 2020 at the Wayback Machine Between 12 January through 31 May 2018, the National IHR Focal Point of The Kingdom of Saudi Arabia reported 75 laboratory confirmed cases of Middle East respiratory syndrome coronavirus (MERS_CoV), including twenty-three (23) deaths. Date www.emro.who.int, accessed 29 January 2020

913) Panetta, Grace. Hong Kong is shutting down its schools until February 17 to limit the spread of the Wuhan coronavirus . Business Insider. Retrieved 25 January 2020.

914) Pang J, Wang MX, Ang IY, Tan SH, Lewis RF, Chen JI, et al. (February 2020). Potential Rapid Diagnostics, Vaccine and Therapeutics for 2019 Novel Coronavirus (2019-nCoV): A Systematic Review . Journal of Clinical Medicine. 9 (3): 623. doi:10.3390/jcm9030623.

915) Papagrigorakis, Manolis J.; Yapijakis, Christos; Synodinos, Philippos N.; Baziotopoulou-Valavani, Effie (2007). DNA examination of ancient dental pulp incriminates typhoid fever as a

probable cause of the Plague of Athens . International Journal of Infectious Diseases. 10 (3): 206–214. doi:10.1016/j.ijid.2005.09.001.

916) Paris, Janella. Philippines confirms 3rd case of novel coronavirus . Rappler.

917) Parker-Pope, Tara (19 March 2020). Deciding How Much Distance You Should Keep . The New York Times. ISSN 0362-4331. Retrieved 20 March 2020.

918) Passengers on cruise liner docked in Japan test positive for coronavirus . The Jerusalem Post | JPost.com. Archived from the original on 6 February 2020. Retrieved 8 February 2020.

919) Past pandemics that ravaged Europe Archived 2017-10-07 at the Wayback Machine, BBC News, 7 November 2005

920) Patiënt met COVID-19 in Delft (in Dutch) . National Institute for Public Health. Retrieved 1 March 2020.

921) Patients (3) get better on arthritis drug . 5 March 2020. Retrieved 14 March 2020.

922) Patients taking ACE-i and ARBs who contract COVID-19 should continue treatment, unless otherwise advised by their physician . Retrieved 21 March 2020.

923) Patterson, K. D.; Pyle, G. F. (1991). The geography and mortality of the 1918 influenza pandemic . Bulletin of the History of Medicine. 65 (1): 4–21.

924) Pearce K (13 March 2020). Antibodies from COVID-19 survivors could be used to treat patients, protect those at risk: Infusions of antibody-laden blood have been used with reported success in prior outbreaks, including the SARS epidemic and the 1918 flu pandemic . The Hub at Johns Hopkins University. Retrieved 14 March 2020.

925) People's Epidemiology Library

926) Pernet E, Downey J, Vinh DC, Powell WS, Divangahi M (August 2019). Leukotriene B4-type I interferon axis regulates macrophage-mediated disease tolerance to influenza infection . Nat Microbiol. 4(8): 1389–1400. doi:10.1038/s41564-019-0444-3.

927) Pfizer and BioNTech announce joint development of a potential COVID-19 vaccine . TechCrunch. 18 March 2020. Retrieved 18 March 2020.

928) PH imposes travel ban on China as new coronavirus infections rise globally . ABS-CBN News. Archived from the original on 5 February 2020. Retrieved 2 February 2020.
929) PH suspends issuance of visa on arrival to Chinese nationals amid coronavirus alarm. CNN Philippines. Retrieved 28 January2020.
930) Physicians work out treatment guidelines for coronavirus . m.koreabiomed.com (in Korean). 13 February 2020.
931) Pickrell, Ryan (2 February 2020). China says Wuhan coronavirus victims who die should be quickly cremated without funerals as death toll rises . Business Insider. Archived from the original on 4 February 2020. Retrieved 8 February 2020.
932) Pickrell, Ryan. Wuhan is scrambling to fill 11 sports centers, exhibition halls, and other local venues with over 10,000 beds to create makeshift coronavirus hospitals . Business Insider. Archived from the original on 5 February 2020. Retrieved 5 February 2020.
933) Plague – Madagascar . WHO. World Health Organization. Archived from the original on 2 May 2019. Retrieved 5 July 2016.
934) Plague in the Democratic Republic of the Congo 13 October 2006 www.who.int, accessed 26 February 2020
935) Plague in the Democratic Republic of the Congo 14 June 2006 www.who.int, accessed 26 February 2020
936) Plague of Athens: Another Medical Mystery Solved at University of Maryland . University of Maryland Medical Center. Archived from the original on 2015-12-04. Retrieved 2016-02-10.
937) PNG to ban all travellers from Asian countries as it steps up response to coronavirus outbreak . ABC News. 28 January 2020. Retrieved 29 January 2020.
938) Pollina, Elvira; Balmer, Crispian (24 February 2020). Third person dies in coronavirus outbreak in northern Italy: official . Reuters. Retrieved 24 February 2020.
939) Post, The Jakarta. Indonesia set to ban live animal imports from China as coronavirus fears grow . The Jakarta Post. Archivedfrom the original on 4 February 2020. Retrieved 3 February 2020.
940) Potter, C. W. (2002). Foreword . Influenza. Elsevier Science. p. vii.

941) Potvrđen šesti slučaj koronavirusa u Hrvatskoj (in Croatian) . Index.hr. Retrieved 28 February 2020.

942) Potvrđen treći slučaj zaraze koronavirusom u Hrvatskoj . N1. Retrieved 26 February 2020.

943) Praveen Duddu. Coronavirus outbreak: Vaccines/drugs in the pipeline for Covid-19 Archived 19 February 2020 at the Wayback Machine. clinicaltrialsarena.com 19 February 2020.

944) Press Trust of India (March 21, 2015). Swine flu deaths at 1895; number of cases near 32K mark . The Indian Express. Archivedfrom the original on October 12, 2015. Retrieved March 21, 2015.

945) Prevention of Coronavirus Disease 2019 (COVID-19) (PDF). Centre for Health Protection. Retrieved 22 March 2020. Wear a surgical mask when taking public transport or staying in crowded places.

946) Prichard, Augustin; Fothergill, John (1894). Influenza in 1775 . The Lancet. 143 (3673): 175–176. doi:10.1016/S0140-6736(01)66026-4. Archived from the original on 2020-01-03. Retrieved 2019-07-05.

947) Prime Minister: 'Prohibit' any flight from Vietnam to the Chinese corona epidemic area . www.xaluan.com. Retrieved 29 January2020.

948) Principles of Epidemiology, Third Edition (PDF). Atlanta, Georgia: Centers for Disease Control and Prevention. 2012.

949) Proper N95 Respirator Use for Respiratory Protection Preparedness. NIOSH Science Blog (16 March 2020). Retrieved 16 March 2020.

950) Pryor, E. G. (1975). The great plague of Hong bong . Journal of the Hong Kong Branch of the Royal Asiatic Society. 15: 61–70. JSTOR 23881624.

951) Purvis, Thomas L. (2014). Colonial America To 1763. Infobase Publishing. p. 173. ISBN 978-1-4381-0799-8.

952) Q&A on coronaviruses (COVID-19) . World Health Organization (WHO). Retrieved 11 March 2020.

953) Q&A on coronaviruses (COVID-19): How long is the incubation period for COVID-19? . World Health Organization

(WHO). Archived from the original on 20 January 2020. Retrieved 26 February 2020.

954) Q&A on coronaviruses . World Health Organization (WHO). Archived from the original on 20 January 2020. Retrieved 27 January 2020.

955) Qatar Air Suspends Flights to China as Cases Jump: Virus Update . Financial Post. 1 February 2020. Archived from the original on 2 February 2020. Retrieved 1 February 2020.

956) Quang Ninh blocks entry of Chinese tourists over nCoV spread threat . Vietnam+. 31 January 2020. Retrieved 31 January 2020.

957) Quảng Ninh đóng cửa các điểm xuất hàng trên tuyến biên giới giáp Trung Quốc (in Vietnamese). Archived from the original on 1 February 2020. Retrieved 31 January 2020.

958) Quebec's first presumptive case of coronavirus detected, health minister says . CBC. 28 February 2020. Retrieved 28 February2020.

959) Railway closures, no visas: Hong Kong scrambles to fight mainland virus outbreak . South China Morning Post. 28 January 2020. Retrieved 28 January 2020.

960) Ramchandani, Nisha (28 January 2020). Straco temporarily shuts 3 China attractions to curb Wuhan virus spread . The Business Times. Retrieved 14 February 2020.

961) Ramsay, Max (15 February 2020). First coronavirus patient dies in Europe . CNN. Retrieved 15 February 2020.

962) Ranlet, Philip (2000). The British, the Indians, and Smallpox: What Actually Happened at Fort Pitt in 1763? . Pennsylvania History: A Journal of Mid-Atlantic Studies. 67 (3): 427–441. ISSN 0031-4528. JSTOR 27774278.

963) Rat Coronavirus—an overview, ScienceDirect Topics . www.sciencedirect.com.

964) Ratcliffe, Rebecca; Standaert, and Michael (27 January 2020). China coronavirus: mayor of Wuhan admits mistakes . The Guardian. ISSN 0261-3077. Archived from the original on 27 January 2020. Retrieved 27 January 2020.

965) Ravenel, Mazÿk P. (May 1938). The Croydon Epidemic of Typhoid Fever . American Journal of Public Health and the Nations Health. 28 (5): 644–646. doi:10.2105/AJPH.28.5.644. PMC 1529192.

966) Rawlinson (now), Kevin; Perraudin, Frances; Walker, Amy; Murphy, Simon; Farrer (earlier), Martin; Gibbs, Samuel; McDonald, Henry (12 February 2020). Coronavirus: first case confirmed in London as major tech event cancelled – latest news . The Guardian. ISSN 0261-3077. Retrieved 12 February 2020.

967) Real-Time RT-PCR Panel for Detection 2019-nCoV . Centers for Disease Control and Prevention. 29 January 2020. Archivedfrom the original on 30 January 2020. Retrieved 1 February 2020.

968) Reanalyzing the 1900–1920 sleeping sickness epidemic in Uganda . Archived from the original on 2009-09-16. Retrieved 2008-01-07.

969) Recovered Covid-19 Patients in Guangdong Tested Positive Again—Caixin Global . www.caixinglobal.com. Retrieved 15 March 2020.

970) Regan, Helen; George, Steve (31 January 2020). Macao delays restarting schools indefinitely . CNN. Retrieved 31 January 2020.

971) Remdesivir Being Evaluated Against Novel Coronavirus . ContagionLive. Archived from the original on 6 February 2020. Retrieved 8 February 2020.

972) Report 4: Severity of 2019-novel coronavirus (nCoV) (PDF). Archived (PDF) from the original on 10 February 2020. Retrieved 10 February 2020.

973) Report of the WHO-China Joint Mission on Coronavirus Disease 2019 (COVID-19) (PDF) (Report). World Health Organization (WHO). 16–24 February 2020. Retrieved 21 March2020.

974) Report sulle caratteristiche dei pazienti deceduti positivi a COVID-19 in Italia (PDF) (Report) (in Italian). Rome: Istituto Superiore di Sanità. 21 March 2020. Retrieved 23 March 2020.

975) Repurposed drugs may help scientists fight the new coronavirus . Science News. 10 March 2020. Retrieved 20 March2020.

976) Roberts M (17 March 2020). Coronavirus: US volunteers test first vaccine . BBC News. Retrieved 17 March 2020.

977) Roberts, Anjeanette; Thomas, William D.; Guarner, Jeannette; Lamirande, Elaine W.; et al. (March 2006). Therapy with a Severe Acute Respiratory Syndrome–Associated Coronavirus–Neutralizing

Human Monoclonal Antibody Reduces Disease Severity and Viral Burden in Golden Syrian Hamsters . The Journal of Infectious Diseases. 193 (5): 685–92. doi:10.1086/500143.

978) Rocca, Ryan (29 February 2020). 3 new COVID-19 cases announced in Ontario, 2 had travel history to Iran . globalnews.ca.

979) Rogers, D.J.; Wilson, A.J.; Hay, S.I.; Graham, A.J. (2006). The Global Distribution of Yellow Fever and Dengue. Advances in Parasitology. 62: 181–220. doi:10.1016/S0065-308X(05)62006-4. ISBN 9780120317622. ISSN 0065-308X.

980) Rohé, George Henry; Robin, Albert (1908). Text-book of Hygiene: A Comprehensive Treatise on the Principles and Practice of Preventive Medicine from an American Standpoint. Davis. p. 428. Retrieved 12 February 2020. spain 1788 dengue fever.

981) Rosen, William (2007), Justinian's Flea: Plague, Empire, and the Birth of Europe Archived 2017-07-24 at the Wayback Machine. Viking Adult; p. 3; ISBN 978-0-670-03855-8.

982) Roser M, Ritchie H, Ortiz-Ospina E (4 March 2020). Coronavirus Disease (COVID-19) . Our World in Data. Retrieved 12 March2020.

983) Rosie (29 January 2020). British Airways Cancels All Flights to Mainland China . Business Insider. Retrieved 29 January 2020.

984) Ross, David. UK travel and heritage – Britain Express UK travel guide . The London Plague of 1665. Archived from the original on 26 May 2019. Retrieved 14 July 2011.

985) RTHK . news.rthk.hk (in Chinese). Archived from the original on 25 January 2020. Retrieved 25 January 2020.

986) Ruan Q, Yang K, Wang W, Jiang L, Song J (March 2020). Clinical predictors of mortality due to COVID-19 based on an analysis of data of 150 patients from Wuhan, China . Intensive Care Medicine. doi:10.1007/s00134-020-05991-x.

987) Russia Limits Rail Service With China From Jan. 31 Over Coronavirus . The New York Times. 30 January 2020.

988) Russia ramps up controls, shuts China border crossings over virus fears . Reuters. 28 January 2020. Retrieved 29 January2020.

989) Russo, Donovan (4 February 2020). Nike says the coronavirus epidemic will have 'material impact' on China operations . CNBC.

Archived from the original on 6 February 2020. Retrieved 5 February 2020.

990) S. Korea adds 2 more virus cases at 18, one confirmed after trip to Singapore . Yonhap. 5 February 2020. Archived from the original on 5 February 2020. Retrieved 5 February 2020.

991) S. Korea adds 3 more virus cases for 27 total, no entry ban expansion . 9 February 2020.

992) S. Korea reports 1 more case of novel coronavirus, total now at 29 . Yonhap. 16 February 2020. Retrieved 16 February 2020.

993) S. Korea reports 1 more case of novel coronavirus, total now at 31 . Yonhap. 18 February 2020. Retrieved 18 February 2020.

994) S. Korea reports 12th confirmed case of novel coronavirus . en.yna.co.kr. Seoul, South Korea: Yonhap News Agency. 1 February 2020. Archived from the original on 1 February 2020. Retrieved 1 February 2020.

995) S. Korea reports 2 more cases of novel coronavirus, total now at 18 . Yonhap. 5 February 2020. Retrieved 5 February 2020.

996) S. Korea reports 2nd death of coronavirus patient . Yonhap. 21 February 2020. Retrieved 21 February 2020.

997) S. Korea reports 3 more cases of novel coronavirus, total now at 15 . Yonhap News Agency. Seoul, South Korea. 2 February 2020. Archived from the original on 2 February 2020. Retrieved 2 February 2020.

998) S. Korea reports 3 more deaths from coronavirus, death toll rises to 16 . Yonhap News Agency. 28 February 2020. Retrieved 28 February 2020.

999) S. Korea reports 334 new virus cases, total now at 1,595 . Yonhap News Agency. 27 February 2020. Retrieved 27 February2020.

1000) Salo, Jackie (24 February 2020). US confirms 53 coronavirus cases, including evacuated cruise passengers . New York Post. Retrieved 25 February 2020.

1001) Sẵn sàng chở công dân Việt Nam 'mắc kẹt' từ Vũ Hán về nước . 27 January 2020. Retrieved 27 January 2020.

1002) Sang-Hun C (2015-06-08). MERS Virus's Path: One Man, Many South Korean Hospitals . The New York Times. Archived from the original on 2017-07-15. Retrieved 2017-03-01.

1003) Sanitat, G. V. A. (27 February 2020). @GVAsanitat confirma 6 nuevos casos de coronavirus en la Comunitat Valenciana, lo que suma un total de 8. La consellera anabarcelochico ha explicado que todos los afectados tienen síntomas leves y la confirmación de estos casos indica que las vías de detección funcionanpic.twitter.com/JoWrbAISjI .

1004) Sanitat, G. V. A. (27 February 2020). @GVAsanitat confirma un segundo caso de coronavirus en la Comunitat Valenciana. Se trata del caso en estudio de un hombre en el @GVAclinic, en Valencia, con síntomas compatibles con la enfermedad. Los primeros análisis han confirmado la presencia del virus .

1005) SARS (severe acute respiratory syndrome) . National Health Service. 5 March 2020. Archived from the original on 9 March 2020. Retrieved 31 January 2020.

1006) Saskatchewan lab joins global effort to develop coronavirus vaccine . CBC News. The Canadian Press. 24 January 2020. Archived from the original on 25 January 2020. Retrieved 25 January 2020.

1007) Scasciamacchia, Silvia; Serrecchia, Luigina; Giangrossi, Luigi; Garofolo, Giuliano; Balestrucci, Antonio; Sammartino, Gilberto; Fasanella, Antonio (2012). Plague Epidemic in the Kingdom of Naples, 1656–1658 – Volume 18, Number 1 – January 2012 – Emerging Infectious Diseases journal – CDC . Emerging Infectious Diseases. 18 (1): 186–8. doi:10.3201/eid1801.110597.

1008) Scoot cancels flights to China's Wuhan over virus outbreak . CNA. 23 January 2020. Retrieved 28 January 2020.

1009) Scott, Dylan (16 March 2020). Coronavirus is exposing all of the weaknesses in the US health system High health care costs and low medical capacity made the US uniquely vulnerable to the coronavirus . Vox. Retrieved 18 March 2020.

1010) Second case of coronavirus detected in Bahrain . Al Arabiya. Retrieved 25 February 2020.

1011) Second coronavirus case in Lebanon: Health ministry . Al Arabiya. Retrieved 26 February 2020.

1012) Second Danish coronavirus infection identified . The Local Denmark. 28 February 2020. Retrieved 28 February 2020.

1013) Second person dies from coronavirus in France: Health official . Al Arabiya. Retrieved 26 February 2020.

1014) Second person-to-person transmission of coronavirus reported in US; 1st death confirmed outside China . 3 February 2020. Archivedfrom the original on 3 February 2020. Retrieved 3 February 2020.

1015) Segundo caso de coronavírus no Brasil é confirmado em SP . R7. 29 February 2020.

1016) Sequence for Putting On Personal Protective Equipment (PPE) (PDF). CDC. Retrieved 8 March 2020.

1017) Severe acute respiratory syndrome coronavirus 2 data hub . NCBI. Retrieved 4 March 2020.

1018) Severe Outcomes Among Patients with Coronavirus Disease 2019 (COVID-19) — United States, February 12–March 16, 2020 . Centers for Disease Control. 16 March 2020. Retrieved 22 March 2020.

1019) Sexton NR, Smith EC, Blanc H, Vignuzzi M, Peersen OB, Denison MR (August 2016). Homology-Based Identification of a Mutation in the Coronavirus RNA-Dependent RNA Polymerase That Confers Resistance to Multiple Mutagens . Journal of Virology. 90 (16): 7415–28. doi:10.1128/JVI.00080-16. PMC 4984655. Finally, these results, combined with those from previous work (33, 44), suggest that CoVs encode at least three proteins involved in fidelity (nsp12-RdRp, nsp14-ExoN, and nsp10), supporting the assembly of a multiprotein replicase-fidelity complex, as described previously (38).

1020) Shahraki, Abdolrazag Hashemi (2016), Plague in Iran: Its history and current status.

1021) Shankland, Stephen. Sixteen supercomputers tackle coronavirus cures in US . CNET. Retrieved 23 March 2020.

1022) Sheahan, Timothy P.; Sims, Amy C.; Leist, Sarah R.; Schäfer, Alexandra; Won, John; Brown, Ariane J.; Montgomery, Stephanie A.; Hogg, Alison; Babusis, Darius; Clarke, Michael O.; Spahn, Jamie E. (10 January 2020). Comparative therapeutic efficacy of remdesivir and combination lopinavir, ritonavir, and interferon beta against

MERS-CoV . Nature Communications. 11 (1): 222. Bibcode:2020NatCo..11..222S. doi:10.1038/s41467-019-13940-6. ISSN 2041-1723. PMC 6954302.
1023) Shehata, M.M., Gomaa, M.R., Ali, M.A. et al. Middle East respiratory syndrome coronavirus: a comprehensive review. Front. Med. 10, 120–136 (2016). doi:10.1007/s11684-016-0430-6
1024) Shortage of personal protective equipment endangering health workers worldwide (Press release). WHO. 3 March 2020. Retrieved 24 March 2020.
1025) Shute, Joe (14 March 2020). The vaccine hunters racing to save the world from the coronavirus pandemic . The Telegraph. Retrieved 18 March 2020.
1026) Sim, Walter (14 February 2020). Coronavirus: 'Stealth outbreak' fears as Japan struggles to find links to new cases nationwide . The Straits Times. Retrieved 15 February 2020.
1027) Sim, Walter (7 February 2020). Coronavirus: Japan reports 41 more positive cases on cruise ship, total cases on Diamond Princess now 61 . The Straits Times. Archived from the original on 7 February 2020. Retrieved 7 February 2020.
1028) Simmons G, Zmora P, Gierer S, Heurich A, Pöhlmann S (December 2013). Proteolytic activation of the SARS-coronavirus spike protein: cutting enzymes at the cutting edge of antiviral research . Antiviral Research. 100 (3): 605–14. doi:10.1016/j.antiviral.2013.09.028.
1029) Singapore households to get 4 face masks each amid worries over Wuhan virus . CNA. Retrieved 30 January 2020.
1030) Singapore Reit closes malls in China amid Wuhan virus spread . The Straits Times. 28 January 2020. Retrieved 14 February 2020.
1031) Singapore turning away new visitors with travel history to mainland China within the last 14 days to contain Wuhan virus .
1032) Situation Report 28 (PDF). World Health Organization. Retrieved 17 February 2020.
1033) Situation Report 29 (PDF). World Health Organization. Retrieved 18 February 2020.
1034) Situation Report 30 (PDF). World Health Organization. Retrieved 19 February 2020.

1035) Situation Report 31 (PDF). World Health Organization. Retrieved 20 February 2020.
1036) Situation Report 32 (PDF). World Health Organization. Retrieved 21 February 2020.
1037) Situation Report 33 (PDF). World Health Organization. Retrieved 22 February 2020.
1038) Situation Report 34 (PDF). World Health Organization. Retrieved 23 February 2020.
1039) Situation Report 35 (PDF). World Health Organization. Retrieved 24 February 2020.
1040) Situation Report 36 (PDF). World Health Organization. Retrieved 25 February 2020.
1041) Situation Report 37 (PDF). World Health Organization. Retrieved 26 February 2020.
1042) Situation Report 38 (PDF). World Health Organization. Retrieved 27 February 2020.
1043) Situation Report 39 (PDF). World Health Organization. Retrieved 28 February 2020.
1044) Situation Report 40 (PDF). World Health Organization. Retrieved 29 February 2020.
1045) Situation summary Latest available situation summary, 26 June 2015. World Health Organisation (2015-06-19). Retrieved on 2015-06-20 . Archived from the original on 5 June 2016. Retrieved 27 June 2015.
1046) Six more people — five in Tokyo, one in Aichi Prefecture — test positive for coronavirus . The Japan Times Online. 16 February 2020. ISSN 0447-5763. Retrieved 16 February 2020.
1047) Sleigh, Sara H.; Barton, Cheryl L. (2010). Repurposing Strategies for Therapeutics . Pharmaceutical Medicine. 24 (3): 151–159. doi:10.1007/BF03256811.
1048) Smallpox decimates tribes; survivors join together – Timeline – Native Voices . www.nlm.nih.gov. Archived from the original on 16 August 2016. Retrieved 4 July 2016.
1049) Smith RD (December 2006). Responding to global infectious disease outbreaks: lessons from SARS on the role of risk perception,

communication and management . Social Science & Medicine. 63 (12): 3113–23. doi:10.1016/j.socscimed.2006.08.004.

1050) Snijder EJ, Bredenbeek PJ, Dobbe JC, Thiel V, Ziebuhr J, Poon LL, et al. (August 2003). Unique and conserved features of genome and proteome of SARS-coronavirus, an early split-off from the coronavirus group 2 lineage . Journal of Molecular Biology. 331 (5): 991–1004. doi:10.1016/S0022-2836(03)00865-9.

1051) Sola I, Alonso S, Zúñiga S, Balasch M, Plana-Durán J, Enjuanes L (April 2003). Engineering the transmissible gastroenteritis virus genome as an expression vector inducing lactogenic immunity . Journal of Virology. 77 (7): 4357–69. doi:10.1128/JVI.77.7.4357-4369.2003.

1052) Somalia cholera death fears grow . 2007-04-28. Archivedfrom the original on 5 October 2017. Retrieved 16 November 2014.

1053) Some 200 Israeli students, teachers ordered into quarantine over virus fears . The Times of Israel. 22 February 2020. Retrieved 23 February 2020.

1054) South Korea coronavirus cases surge to 2,337, BTS cancel gigs . The Straits Times. 28 February 2020. Retrieved 29 February 2020.

1055) South Korea COVID-19 cases exceeds 1,200; US soldier among new cases . CNA. 26 February 2020. Retrieved 26 February2020.

1056) South Korea reports 229 more coronavirus cases, bringing total to 433 . The Straits Times. 22 February 2020. Retrieved 22 February 2020.

1057) South Korea reports 594 more coronavirus cases, total 2,931 . CNA. 29 February 2020. Retrieved 29 February 2020.

1058) South Korea reports first COVID-19 death as infected cases reach 104 . CNA. 20 February 2020. Retrieved 20 February 2020.

1059) South Korean coronavirus cases jump by two-thirds in one day; church services seen as centre of spike . The Straits Times. 19 February 2020. Retrieved 19 February 2020.

1060) Spain confirms first case of coronavirus . The Straits Times. 1 February 2020. Retrieved 26 February 2020.

1061) Spain detects 10 new cases of coronavirus in 36 hours . El Pais. Retrieved 26 February 2020.

1062) Spain seeks to calm nerves as COVID-19 cases jump . CNA. 26 February 2020. Retrieved 26 February 2020.

1063) Spanish authorities confirm Briton is country's second coronavirus case . Reuters. 9 February 2020. Retrieved 9 February 2020.

1064) Sparrow A. How China's Coronavirus Is Spreading—and How to Stop It . Foreign Policy. Archived from the original on 31 January 2020. Retrieved 2 February 2020.

1065) SPIEGEL, DER. Lungenkrankheit: Zahl der Coronavirus-Fälle in Deutschland steigt auf 14 - DER SPIEGEL - Wissenschaft . spiegel.de.

1066) Spinney, Laura (18 March 2020). When will a coronavirus vaccine be ready? . The Guardian. Retrieved 18 March 2020.

1067) Sri Lanka suspends visa on arrival for Chinese citizens after first case of coronavirus . The Hindu. 28 January 2020. Retrieved 28 January 2020.

1068) Sriring, Orathai; Thepgummapat, Panarat; Sarkar, Himani (15 February 2020). Thailand reports one new case of coronavirus, brings total to 34 . Reuters. Retrieved 15 February 2020.

1069) Staff Reporter. Cholera epidemic takes hold in Senegal . The M&G Online. Archived from the original on 3 February 2017. Retrieved 16 November 2014.

1070) Stanway, David (29 January 2020). China Police Drop Case of SARS 'Rumour-Mongering' at Start of New Crisis . The New York Times. Archived from the original on 30 January 2020. Retrieved 30 January 2020.

1071) State Department authorizes personnel to evacuate China due to coronavirus . Axios.

1072) Statement from the National Public Health Emergency Team . BNO News. 29 February 2020. Retrieved 29 February 2020.

1073) Statement on the second meeting of the International Health Regulations (2005) Emergency Committee regarding the outbreak of novel coronavirus (2019-nCoV) . World Health Organization(WHO). Archived from the original on 31 January 2020. Retrieved 11 February 2020.

1074) Statement on the second meeting of the International Health Regulations (2005) Emergency Committee regarding the outbreak of novel coronavirus (2019-nCoV) . www.who.int.

1075) Steenhuysen J, Kelland K (24 January 2020). With Wuhan virus genetic code in hand, scientists begin work on a vaccine . Reuters. Archived from the original on 25 January 2020. Retrieved 25 January 2020.

1076) Steenhuysen J, Kelland K (24 January 2020). With Wuhan virus genetic code in hand, scientists begin work on a vaccine . Reuters. Archived from the original on 25 January 2020. Retrieved 25 January 2020.

1077) Strategies for Optimizing the Supply of Eye Protection . CDC.

1078) Strategies for Optimizing the Supply of Facemasks . CDC.

1079) Strong, Matthew (25 February 2020). Taiwan's coronavirus case No. 31 is 11-year-old grandson of a previous patient . Taiwan News. Retrieved 26 February 2020.

1080) Strong, Matthew (26 February 2020). Foreign caregiver announced as Taiwan's coronavirus case No. 32 . Taiwan News. Retrieved 26 February 2020.

1081) Sturman LS, Holmes KV (1983). Lauffer MA, Maramorosch K (eds.). The molecular biology of coronaviruses . Advances in Virus Research. 28: 35–112. doi:10.1016/s0065-3527(08)60721-6. These viruses displayed a characteristic fringe of large, distinctive, petal-shaped peplomers or spikes which resembled a crown, like the corona spinarum in religious art; hence the name coronaviruses.

1082) Subscribe to read | Financial Times . www.ft.com. Archivedfrom the original on 4 February 2020. Retrieved 8 February 2020.

1083) Sugiyama, Satoshi (12 February 2020). Japan to expand coronavirus testing as 39 more on cruise test positive . The Japan Times Online. Retrieved 13 February 2020.

1084) Sum, Lok-kei; Chan, Ho-him; Chung, Kimmy (11 February 2020). Coronavirus: seven more cases confirmed in Hong Kong, taking city's total number to 49 . South China Morning Post. Retrieved 11 February 2020.

1085) Sum, Lok-kei; Leung, Kanis; Cheung, Elizabeth; Leung, Christy; Lo, Clifford (20 February 2020). Coronavirus: officer becomes 69th case in Hong Kong after attending party with 59 others, all of whom have been asked to avoid public . South China Morning Post. Retrieved 21 February 2020.

1086) Sum, Lok-kei; Lum, Alvin; Chan, Ho-him (16 February 2020). Coronavirus: husband and wife hospitalised after man becomes Hong Kong's 57th confirmed patient . South China Morning Post. Retrieved 16 February 2020.

1087) Sum, Lok-kei; Mok, Danny (11 February 2020). Coronavirus: More than 100 Tsing Yi residents evacuated as Hong Kong health authorities confirm two cases from same block . South China Morning Post. Retrieved 11 February 2020.

1088) Surveillance, forecasting and response International conference on dengue control, 27–29 February 2012 www.emro.who.intaccessed 16 February 2020

1089) Suzuki, A. (2011). Smallpox and the epidemiological heritage of modern Japan: Towards a total history . Medical History. 55 (3): 313–18. doi:10.1017/S0025727300005329.

1090) Swanky, Tom (2016). The Smallpox War in Nuxalk Territory. British Columbia, Canada: Dragon Heart. p. 93. ISBN 978-1365410161.

1091) Swilam, Alaa; Ghantous, Ghaida; Cornwell, Alexander; Russell, Ros (22 February 2020). UAE records two new coronavirus cases, total number reaches 11 . Reuters. Retrieved 22 February 2020.

1092) Switzerland confirms sixth coronavirus case . swissinfo.ch. Retrieved 27 February 2020.

1093) Symptoms of Novel Coronavirus (2019-nCoV) . www.cdc.gov. 10 February 2020. Archived from the original on 30 January 2020. Retrieved 11 February 2020.

1094) tagesschau.de. Nach Flug aus China: Zwei Corona-Infizierte in Südpfalz-Kaserne . tagesschau.de (in German). Archived from the original on 3 February 2020. Retrieved 7 February 2020.

1095) Taiwan Centers for Disease Control. 28 February 2020. Retrieved 28 February 2020.

1096) Taiwan confirms first COVID-19 death . CNA. 16 February 2020. Retrieved 16 February 2020.
1097) Taiwan restricts travellers from Hong Kong and Macau amid crisis . South China Morning Post. 6 February 2020. Archivedfrom the original on 6 February 2020. Retrieved 8 February 2020.
1098) Taiwan slaps coronavirus entry ban on foreigners from mainland China . South China Morning Post. 4 February 2020. Archivedfrom the original on 6 February 2020. Retrieved 8 February 2020.
1099) Taiwan's confirms first coronavirus death on island, cases at 20 . Reuters. 16 February 2020. Retrieved 16 February 2020.
1100) Taiwan's COVID-19 cases rise to 28 . Taiwan News. 23 February 2020. Retrieved 24 February 2020.
1101) Tajima M (1970). Morphology of transmissible gastroenteritis virus of pigs. A possible member of coronaviruses. Brief report . Archiv für die Gesamte Virusforschung. 29 (1): 105–8. doi:10.1007/BF01253886.
1102) Takis, a biotech company in Castel Romano, Rome, announced that it is ready to test its COVID-19 vaccine on pre-clinical models (PDF) (Press release). Rome: Takis Biotech. 17 March 2020. Retrieved 24 March 2020.
1103) Tan, Weizhen (12 February 2020). Coronavirus live updates: China's Hubei reports 14,840 new cases, 242 additional deaths . CNBC. Retrieved 13 February 2020.
1104) Tan, Wenjie; Zhao, Xiang; Ma, Xuejun; Wang, Wenling; Niu, Peihua; Xu, Wenbo; Gao, George F.; Wu, Guizhen (1 January 2020). A Novel Coronavirus Genome Identified in a Cluster of Pneumonia Cases — Wuhan, China 2019–2020 . China CDC Weekly. 2 (4): 61–62. Archived from the original on 7 February 2020. Retrieved 8 February 2020 – via weekly.chinacdc.cn.
1105) Tashny, Sukumaran (6 February 2020). Coronavirus: Malaysia confirms first case of human transmission from Singaporte meeting, taking tally to 14 . South China Morning Post. South China Morning Post. Archived from the original on 6 February 2020. Retrieved 7 February 2020.

1106) Tate, Curtis. Coronavirus outbreak: United, American suspend flights to and from Hong Kong . USA TODAY. Archived from the original on 6 February 2020. Retrieved 5 February 2020.

1107) Taylor, Marisa. (22 March 2020) Exclusive: U.S. axed CDC expert job in China months before virus outbreak . Reuters websiteRetrieved 23 March 2020.

1108) Team, CM1-US. Four suspected cases of mystery Chinese viral pneumonia detected at CM Airport | Chiang Mai One . Archivedfrom the original on 15 January 2020. Retrieved 15 January 2020.

1109) Tee, Zhuo (16 February 2020). Coronavirus: SAF regular among three new cases confirmed in Singapore . The Straits Times. Retrieved 16 February 2020.

1110) Tee, Zhuo; Goh, Timothy (24 February 2020). Coronavirus: 1 new case in S'pore, a 75-year-old woman linked to Life Church and Missions; 2 others discharged . The Straits Times. Retrieved 24 February 2020.

1111) Ten confirmed infected on ship near Yokohama . NHK World. 4 February 2020. Archived from the original on 5 February 2020. Retrieved 4 February 2020.

1112) Tencent (6 February 2020). The Tencent News Epidemic Situation Tracker reports real-time data from China' s National Health Commission and various Municipal Health Commissions across China . @TencentGlobal. Retrieved 8 February 2020.

1113) Teng, Amelia (23 January 2020). Wuhan virus: Singapore schools ask parents, staff to declare weekend travel plans . The Straits Times. Retrieved 29 January 2020.

1114) Tercer cas importat lleu de coronavirus a Catalunya . Departament de Salut. Retrieved 27 February 2020.

1115) Tesla tumbles as coronavirus weighs on Shanghai factory . Reuters. 5 February 2020. Archived from the original on 6 February 2020. Retrieved 5 February 2020.

1116) Texas Department of State Health Services, History of Plague . Archived from the original on 2013-06-26. Retrieved 2008-07-01.

1117) Thaiger, The (9 January 2020). Chinese scientists identify the 'Wuhan Virus'. Screening continues on Thai-bound flights . The

Thaiger. Archived from the original on 10 January 2020. Retrieved 8 February 2020.

1118) Thailand confirms 6 more Wuhan virus infections, bringing total to 14 . CNA. 28 January 2020. Retrieved 28 January 2020.

1119) Thailand confirms 6 new cases of novel coronavirus, including 4 Thais . CNA. 4 February 2020. Archived from the original on 4 February 2020. Retrieved 4 February 2020.

1120) Thailand has 3 new coronavirus cases, urges travel disclosure . Bangkok Post. 26 February 2020. Retrieved 26 February 2020.

1121) Thailand reports 1 new case of coronavirus, brings total to 33 . Bangkok Post. 11 February 2020. Retrieved 11 February 2020.

1122) Thailand reports one new coronavirus case, total reaches 41 - health official . Reuters. 28 February 2020. Retrieved 28 February2020.

1123) Thailand reports seven new coronavirus cases - health ministry . National Post. 8 February 2020. Retrieved 8 February2020.

1124) Thailand reports two new COVID-19 cases, bringing total to 37 . CNA. 25 February 2020. Retrieved 25 February 2020.

1125) The control and eradication of smallpox in South Asia . www.smallpoxhistory.ucl.ac.uk. Archived from the original on 2008-10-19. Retrieved 2008-12-18.

1126) The coronavirus spreads racism against—and among—ethnic Chinese . The Economist. 17 February 2020. Archived from the original on 17 February 2020. Retrieved 17 February 2020.

1127) The Editorial Board (2020-01-29). Is the World Ready for the Coronavirus?—Distrust in science and institutions could be a major problem if the outbreak worsens . The New York Times. Retrieved 2020-01-30.

1128) The Encyclopedia of Ecology and Environmental Management, Blackwell Science. Oxford: Blackwell Publishers, 1998. Credo Reference. Web. 17 September 2012.

1129) The first COVID-19 case originated on November 17, according to Chinese officials searching for 'Patient Zero' . www.msn.com. Retrieved 13 March 2020.

1130) The History of Dengue Outbreaks in the Americaswww.ncbi.nlm.nih.gov, accessed 16 February 2020

1131) The History of Small-Pox in Australia, 1788–1908, JHL Cumpston, (1914, Government Printer, Melb.)This epidemic is unlikely to have been a natural event. see, Warren (2013) doi:10.1080/14443058.2013.849750 After Cook and coinciding with Colonisation With the arrival of the Europeans, the Gadigal population was virtually wiped. In 1789 and 1790 a smallpox epidemic swept through the Aboriginal population around Sydney Archived 2008-06-25 at the Wayback Machine

1132) The Latest: All airports and ports have been closed in order to contain the virus. Shanghai has exploded . ABC News. Retrieved 31 January 2020.

1133) The Ministry of Public Health announces the registration of the first confirmed case of coronavirus 2019 (Covid-19) in the country . Twitter. Qatar News Agency. 29 February 2020. Retrieved 29 February 2020.

1134) The Ministry_of_Health announces the registration of a new novel Corona virus (COVID-19) case, linked to travelling to Iran . Twitter. Oman Ministry of Health. Retrieved 27 February 2020.

1135) The Next Epidemic — Lessons from Ebola , essay by Bill Gates, New England Journal of Medicine (March 18, 2015).

1136) Thêm 2 người Vĩnh Phúc dương tính nCoV - VnExpress Sức Khỏe . vnexpress.net. Archived from the original on 6 February 2020. Retrieved 8 February 2020.

1137) Thêm một người Vĩnh Phúc dương tính nCoV . VnExpress. 9 February 2020. Retrieved 9 February 2020.

1138) They survived the coronavirus. Then they tested positive again. Why? . Los Angeles Times. 13 March 2020. Retrieved 15 March2020.

1139) Thiel V (editor). (2007). Coronaviruses: Molecular and Cellular Biology (1st ed.). Caister Academic Press. ISBN 978-1-904455-16-5.

1140) Third case of coronavirus infection spotted in Finland . Daily Finland. 28 February 2020. Retrieved 28 February 2020.

1141) Thompson, Nicole (23 February 2020). New presumptive case of coronavirus in Toronto after previous three cleared of COVID-19 . National Post. Retrieved 24 February 2020.

1142) Thorp, H. Holden (23 March 2020). Underpromise, overdeliver (PDF). American Association for the Advancement of Science. Retrieved 23 March 2020.
1143) Thousands hit by Brazil outbreak of dengue edition.cnn.com, accessed 16 February 2020
1144) Three more Iran travelers infected with coronavirus in Bahrain, bringing total to 26 . Al Arabiya. 26 February 2020. Retrieved 26 February 2020.
1145) Three Russians evacuated from Diamond Princess diagnosed with coronavirus in Kazan . meduza.io. Retrieved 26 February2020.
1146) Thủ tướng: Sẵn sàng tuyên bố tình trạng khẩn cấp y tế về dịch virus corona . tuoitre.vn (in Vietnamese). 30 January 2020. Retrieved 30 January 2020.
1147) Tian, Dai-Shi; Wang, Wei; Shang, Ke; Ma, Ke; Xie, Cuihong; Tao, Yu; Yang, Sheng; Zhang, Shuoqi; Hu, Ziwei; Zhou, Luoqi; Qin, Chuan (12 March 2020). Dysregulation of immune response in patients with COVID-19 in Wuhan, China . Clinical Infectious Diseases. doi:10.1093/cid/ciaa248.
1148) Tibetan New Year activities called off amid virus outbreak - Xinhua | English.news.cn . www.xinhuanet.com. Retrieved 12 February 2020.
1149) Tidy, Joe (17 March 2020). Coronavirus: Israel enables emergency spy powers . BBC News. Retrieved 18 March 2020.
1150) Tiempo, Casa Editorial El (4 February 2020). Así es la prueba que hará Colombia para detectar el coronavirus . El Tiempo.
1151) Tiger mosquitoes and the history of yellow fever and dengue in Spain . Archived from the original on 4 August 2017. Retrieved 16 November 2014.
1152) Times, The New York (3 February 2020). Hong Kong Reports First Death From Coronavirus Outbreak . Archived from the original on 6 February 2020. Retrieved 8 February 2020 – via NYTimes.com.
1153) Timothy Brook (1999). The Confusions of Pleasure: Commerce and Culture in Ming China. University of California Press. p. 163. ISBN 978-0-520-22154-3.

1154) Ting, Victor (28 February 2020). Coronavirus: Hong Kong publishes list of buildings housing quarantined people . South China Morning Post. Retrieved 29 February 2020.
1155) Tirotta E, Carbajal KS, Schaumburg CS, Whitman L, Lane TE (July 2010). Cell replacement therapies to promote remyelination in a viral model of demyelination . Journal of Neuroimmunology. 224(1–2): 101–07. doi:10.1016/j.jneuroim.2010.05.013.
1156) Tjendro, Johannes (13 February 2020). Malaysia confirms new COVID-19 case related to two others who had tested positive . CNA. Retrieved 13 February 2020.
1157) Tognotti, Eugenia (February 2013). Lessons from the History of Quarantine, from Plague to Influenza A . Emerging Infectious Diseases. 19 (2): 254–259. doi:10.3201/eid1902.120312.
1158) Toh, Ting Wei; Kurohi, Rei (27 February 2020). Coronavirus: 12-year-old RI student and his family member among 3 new confirmed cases in S'pore; 4 others discharged . The Straits Times. Retrieved 27 February 2020.
1159) Tokyo taxi driver and Wakayama surgeon test positive for COVID-19 . The Japan Times Online. 13 February 2020. Retrieved 13 February 2020.
1160) Top political advisory body considers delaying annual session in March . news.cgtn.com. Retrieved 17 February 2020.
1161) Total confirmed cases of COVID-19 per million people . Our World in Data. Retrieved 20 March 2020.
1162) Total confirmed deaths due to COVID-19 per million people . Our World in Data. Retrieved 24 March 2020.
1163) Touret F, de Lamballerie X (March 2020). Of chloroquine and COVID-19. Antiviral Research. 177: 104762. doi:10.1016/j.antiviral.2020.104762.
1164) TP.HCM xây dựng 2 bệnh viện dã chiến quy mô 500 giường phòng dịch corona . tuoitre.vn (in Vietnamese). Archived from the original on 4 February 2020. Retrieved 3 February 2020.
1165) Tracking coronavirus: Map, data and timeline . BNO News. 18 February 2020. Retrieved 20 February 2020.
1166) Tracking coronavirus: Map, data and timeline . BNO News. 26 February 2020. Retrieved 26 February 2020.

1167) Tracking coronavirus: Map, data and timeline . BNO News. Retrieved 20 February 2020.
1168) Tracking coronavirus: Map, data and timeline . BNO News. Retrieved 9 February 2020.
1169) Tracking coronavirus: Map, data and timeline . BNO News. Retrieved 9 February 2020.
1170) Tracking coronavirus: Map, data and timeline . Bnonews. bnonews.com. 2 February 2020. Retrieved 16 February 2020.
1171) Tracking the source of New Coronavirus gene sequencing: when the alarm sounded] (in Chinese). Caixin. 26 February 2020. Archived from the original on 27 February 2020. Retrieved 1 March 2020.
1172) Transmission of Novel Coronavirus (2019-nCoV) | CDC . www.cdc.gov. 2020-01-31. Retrieved 2020-02-01.
1173) transmission. Saunders Comprehensive Veterinary Dictionary. Philadelphia: Elsevier Health Sciences, 2007. Credo Reference. Web. 17 September 2012.
1174) Travel advice and advisories for China . 29 January 2020. Retrieved 29 January 2020.
1175) Tripp, Ralph A.; Haynes, Lia M.; Moore, Deborah; Anderson, Barbara; et al. (September 2005). Monoclonal antibodies to SARS-associated coronavirus (SARS-CoV): Identification of neutralizing and antibodies reactive to S, N, M and E viral proteins . Journal of Virological Methods. 128 (1–2): 21–8. doi:10.1016/j.jviromet.2005.03.021.
1176) Trump declares coronavirus outbreak a public emergency, will ban foreign travel from China and quarantine US citizens . Washington Examiner. 31 January 2020.
1177) Turkish Airlines says it suspends flights to China until February 9 . Reuters. 30 January 2020 – via www.reuters.com.
1178) Turner, David (November 1990). The Politics of Despair: The Plague of 746–747 and Iconoclasm in the Byzantine Empire1 . Annual of the British School at Athens. 85: 419–434. doi:10.1017/S006824540001577X. ISSN 2045-2403.

1179) Tweede patiënt in Nederland met COVID-19 (in Dutch) . Ministerie van Volksgezondheid, Welzijn en Sport. 28 February 2020. Retrieved 28 February 2020.
1180) Two elderly in Japan dies after catching COVID-19 . CNA. 26 February 2020. Retrieved 27 February 2020.
1181) Two more cases of coronavirus confirmed in Romania . Romania Insider. 28 February 2020. Retrieved 28 February 2020.
1182) Two more deaths from measles in samoa over new year period . Radio New Zealand. 2020-01-07. Archived from the original on 2020-01-07.
1183) Two new cases of coronavirus in UAE . gulfnews.com.
1184) Two new cases of coronavirus reported in Kuwait: Ministry of Health . Al Arabiya. 28 February 2020. Retrieved 28 February2020.
1185) Two new coronavirus cases in Singapore, including Certis Cisco employee who served quarantine orders . CNA. 10 February 2020. Retrieved 10 February 2020.
1186) U.S. woman previously on Westerdam cruise ship tests positive for new coronavirus . Global News. 15 February 2020. Retrieved 15 February 2020.
1187) UAE confirms recovery of two coronavirus patients, registers six new cases . Al Arabiya. Retrieved 27 February 2020.
1188) UAE Reports 9th Case of Coronavirus Infection | Indiablooms - First Portal on Digital News Management .
1189) UAE's fifth coronavirus case: man's family tests negative . The National. Archived from the original on 5 February 2020. Retrieved 8 February 2020.
1190) UH one of first in nation to begin clinical trials for drug to help patients with COVID-19 . 20 March 2020.
1191) UN health chief announces global 'solidarity trial' to jumpstart search for COVID-19 treatment . UN News. 18 March 2020. Retrieved 23 March 2020.
1192) UNAIDS (2010) report on the global AIDS epidemic'
1193) Vaccine-linked polio hits Nigeria Archived 2016-08-20 at the Wayback Machine, BBC News
1194) Vågene, Åshild J.; Herbig, Alexander; Campana, Michael G.; Robles García, Nelly M.; Warinner, Christina; Sabin, Susanna;

Spyrou, Maria A.; Andrades Valtueña, Aida; Huson, Daniel; Tuross, Noreen; Bos, Kirsten I.; Krause, Johannes (2018). Salmonella enterica genomes from victims of a major sixteenth-century epidemic in Mexico . Nature Ecology & Evolution. 2 (3): 520–528. doi:10.1038/s41559-017-0446-6.

1195) Vaxart (VXRT) - A long shot or perfect shot? . NASDAQ, RTTNews.com. 25 February 2020. Retrieved 1 March 2020.

1196) Velavan TP, Meyer CG (March 2020). The COVID-19 epidemic . Tropical Medicine & International Health. n/a (n/a): 278–80. doi:10.1111/tmi.13383.

1197) Venezuela imposes entry restrictions over coronavirus . Prensa Latina (English). 3 February 2019. Retrieved 4 February 2019.

1198) Vetter P, Eckerle I, Kaiser L (February 2020). Covid-19: a puzzle with many missing pieces. BMJ. 368: m627. doi:10.1136/bmj.m627.

1199) Video Discussion of the Prostate Cancer Epidemic

1200) Việt Nam News. Việt Nam confirms 13th case of coronavirus . Việt Nam News. Việt Nam News. Archived from the original on 7 February 2020. Retrieved 7 February 2020.

1201) Việt Nam tạm thời không cấp thị thực du lịch cho người đến từ vùng dịch viêm phổi Vũ Hán . thanhnien.vn (in Vietnamese). 29 January 2020. Retrieved 30 January 2020.

1202) Vietnam aviation authority ceases all flights to and from coronavirus-stricken Wuhan . tuoitrenews.vn. 24 January 2020. Retrieved 28 January 2020.

1203) Vietnam declares novel coronavirus epidemic . Vietnam+. February 2020. Archived from the original on 1 February 2020. Retrieved 1 February 2020.

1204) Vietnam measles outbreak kills more than 100 people, mostly children . Sydney Morning Herald. 18 April 2014.

1205) Vietnam on alert as common virus kills 81 children – Yahoo News Archived 2016-03-05 at the Wayback Machine. News.yahoo.com (2011-08-19). Retrieved on 2014-05-12.

1206) Vietnam PM urges action against diarrhea outbreak Archived2008-04-16 at the Wayback Machine, Thanh Nien Daily

1207) Vietnam to fly last Wuhan visitors home . Retrieved 27 January2020.
1208) Vijaykrishna D, Smith GJ, Zhang JX, Peiris JS, Chen H, Guan Y (April 2007). Evolutionary insights into the ecology of coronaviruses . Journal of Virology. 81 (8): 4012–20. doi:10.1128/jvi.02605-06.
1209) Vijgen L, Keyaerts E, Moës E, Thoelen I, Wollants E, Lemey P, et al. (February 2005). Complete genomic sequence of human coronavirus OC43: molecular clock analysis suggests a relatively recent zoonotic coronavirus transmission event . Journal of Virology. 79 (3): 1595–604. doi:10.1128/jvi.79.3.1595-1604.2005.
1210) Virology: Coronaviruses . Nature. 220 (5168): 650–650. 1968. doi:10.1038/220650b0. ISSN 0028-0836.
1211) Virus Taxonomy: 2018b Release . International Committee on Taxonomy of Viruses (ICTV). March 2019. Archived from the original on 2018-03-04. Retrieved 2020-01-24.
1212) Vivian, Steve; Zwartz, Henry (21 February 2020). Coronavirus cases confirmed among Australian Diamond Princess evacuees in Darwin . ABC News (Australia). Retrieved 21 February 2020.
1213) voachinese.com (in Chinese). Archived from the original on 30 January 2020. Retrieved 30 January 2020.
1214) Vogel, Gretchen (2020). New blood tests for antibodies could show true scale of coronavirus pandemic . Science. doi:10.1126/science.abb8028. ISSN 0036-8075.
1215) Vulnerable Pacific islands take hard line on coronavirus . Retrieved 4 February 2020.
1216) Wakatsuki, Yoko; Sidhu, Sandi; Krever, Mick (15 February 2020). Diamond Princess cruise ship has 67 more coronavirus cases . CNN. Retrieved 15 February 2020.
1217) Waldman, Carl; Braun, Molly (2009). Atlas of the North American Indian. Infobase Publishing. p. 295. ISBN 978-1-4381-2671-5. Retrieved 12 February 2020.
1218) Walker, James (14 March 2020). Traces Coronavirus to First Confirmed Case, Nearly Identifying 'Patient Zero' . Newsweek. Retrieved 14 March 2020.

1219) Wang C, Horby PW, Hayden FG, Gao GF (February 2020). A novel coronavirus outbreak of global health concern . Lancet. 395(10223): 470–473. doi:10.1016/S0140-6736(20)30185-9.

1220) Wang M, Cao R, Zhang L, Yang X, Liu J, Xu M, et al. (February 2020). Remdesivir and chloroquine effectively inhibit the recently emerged novel coronavirus (2019-nCoV) in vitro . Cell Research. 30 (3): 269–71. doi:10.1038/s41422-020-0282-0.

1221) Wang T, Du Z, Zhu F, Cao Z, An Y, Gao Y, Jiang B (March 2020). Comorbidities and multi-organ injuries in the treatment of COVID-19 . Lancet. Elsevier BV. 395 (10228): e52. doi:10.1016/s0140-6736(20)30558-4.

1222) Wang V (5 March 2020). Most Coronavirus Cases Are Mild. That's Good and Bad News . The New York Times.

1223) Wang W, Tang J, Wei F (April 2020). Updated understanding of the outbreak of 2019 novel coronavirus (2019-nCoV) in Wuhan, China . Journal of Medical Virology. 92 (4): 441–47. doi:10.1002/jmv.25689.

1224) Wang, Manli; Cao, Ruiyuan; Zhang, Leike; Yang, Xinglou; Liu, Jia; Xu, Mingyue; Shi, Zhengli; Hu, Zhihong; Zhong, Wu; Xiao, Gengfu (March 2020). Remdesivir and chloroquine effectively inhibit the recently emerged novel coronavirus (2019-nCoV) in vitro . Cell Research. 30 (3): 269–271. doi:10.1038/s41422-020-0282-0. ISSN 1748-7838.

1225) Water Transmission and COVID-19 (CDC, accessed 19 March 2020)

1226) Weaver, Damien Gayle (now) Matthew; Murray, Jessica; Rourke (earlier), Alison; Murphy, Simon; McCurry, Justin (30 January 2020). Coronavirus: health chief in Chinese city near Wuhan sacked – live news . The Guardian. ISSN 0261-3077. Retrieved 30 January2020.

1227) Wei X, She G, Wu T, Xue C, Cao Y (February 2020). PEDV enters cells through clathrin-, caveolae-, and lipid raft-mediated endocytosis and traffics via the endo-/lysosome pathway . Veterinary Research. 51 (1): 10. doi:10.1186/s13567-020-0739-7. PMC 7011528.

1228) Wei, Xiao-Shan; Wang, Xuan; Niu, Yi-Ran; Ye, Lin-Lin; Peng, Wen-Bei; Wang, Zi-Hao; Yang, Wei-Bing; Yang, Bo-Han; Zhang, Jian-Chu; Ma, Wan-Li; Wang, Xiao-Rong (26 February 2020). Clinical Characteristics of SARS-CoV-2 Infected Pneumonia with Diarrhea. Rochester, NY. SSRN 3546120 Check |ssrn= value (help).

1229) Weiss SR, Navas-Martin S (December 2005). Coronavirus pathogenesis and the emerging pathogen severe acute respiratory syndrome coronavirus. Microbiology and Molecular Biology Reviews. 69 (4): 635–64. doi:10.1128/MMBR.69.4.635-664.2005.

1230) Wertheim JO, Chu DK, Peiris JS, Kosakovsky Pond SL, Poon LL (June 2013). A case for the ancient origin of coronaviruses . Journal of Virology. 87 (12): 7039–45. doi:10.1128/JVI.03273-12.

1231) What are the prospects for a Covid-19 treatment? . The Guardian. 19 March 2020.

1232) When and how to use masks. World Health Organization(WHO). Retrieved 8 March 2020.

1233) WHO – Chikungunya in India. Archived from the original on 23 December 2013. Retrieved 16 November 2014.

1234) WHO Director-General's opening remarks at the media briefing on COVID-19 . World Health Organization (WHO) (Press release). 11 March 2020. Retrieved 12 March 2020.

1235) WHO Director-General's opening remarks at the mission briefing on COVID-19 - 26 February 2020 .

1236) WHO Director-General's statement on the advice of the IHR Emergency Committee on Novel Coronavirus . World Health Organization (WHO).

1237) WHO probes Singapore conference linked to coronavirus spread . The Straits Times. 7 February 2020. Archived from the original on 6 February 2020. Retrieved 8 February 2020.

1238) WHO Regional Office for the Eastern Mediterranean Cholera Situation in Yemen November 2019 . Retrieved 30 November 2019.

1239) WHO says 2019 'worse than usual' for dengue in Pacific . Radio New Zealand. 26 August 2019.

1240) WHO Statement Regarding Cluster of Pneumonia Cases in Wuhan, China . www.who.int. 2020-01-09. Archived from the original on 2020-01-14. Retrieved 2020-01-10.

1241) WHO, Zika: the origin and spread of a mosquito-borne virus . WHO. Retrieved 19 February 2020.

1242) WHO: 'no known effective' treatments for new coronavirus . Reuters. 5 February 2020. Archived from the original on 6 February 2020. Retrieved 6 February 2020.

1243) WHO; NHK. 24 January 2020. Archived from the original on 26 January 2020. Retrieved 3 February 2020.

1244) WHO-recommended handrub formulations . WHO Guidelines on Hand Hygiene in Health Care: First Global Patient Safety Challenge Clean Care Is Safer Care. World Health Organization. 19 March 2009. Retrieved 19 March 2020.

1245) Why Cholera Persists in Haiti Despite an Abundance of Aid . www.npr.org. Archived from the original on 2015-04-28. Retrieved 2018-04-05.

1246) Why Dengue Fever Cases Are Hitting Record Highs In Latin America . NPR. 19 December 2019.

1247) Widegren H, Andersson M, Borgeat P, Flamand L, Johnston S, Greiff L (July 2011). LTB4 increases nasal neutrophil activity and conditions neutrophils to exert antiviral effects . Respir Med. 105 (7): 997–1006. doi:10.1016/j.rmed.2010.12.021.

1248) Wife of first WA coronavirus patient has also tested positive . 7News. 28 February 2020. Retrieved 28 February 2020.

1249) Wiles S (9 March 2020). The three phases of Covid-19 – and how we can make it manageable . The Spinoff. Retrieved 9 March2020.

1250) Wiles, Siouxsie (14 March 2020). After 'Flatten the Curve', we must now 'Stop the Spread'. Here's what that means . The Spinoff. Retrieved 13 March 2020.

1251) Wilford, John Noble (15 April 2008). How Epidemics Helped Shape the Modern Metropolis . The New York Times. Retrieved 13 February 2020.

1252) Will take one-and-a-half to two years for India to develop vaccine for COVID-19: Health Ministry . Economic Times. 12 March 2020. Retrieved 12 March 2020.

1253) William E. Paul (1 May 2008). Fundamental immunology. Lippincott Williams & Wilkins. ISBN 978-0-7817-6519-0. Archivedfrom the original on 2 July 2014. Retrieved 29 March 2011.

1254) Williams, Ann (29 January 2020). CapitaLand, more Singapore companies suspend some businesses in China amid Wuhan virus spread . The Straits Times. Retrieved 16 February 2020.

1255) Wilson, Joanna (20 March 2020). In pictures: the Imperial lab developing a COVID-19 vaccine (Press release). London: Imperial College London. Retrieved 24 March 2020.

1256) Woman from Wuhan is latest coronavirus case in Malaysia, bringing total to 15 . The Straits Times. The Straits Times. 7 February 2020. Retrieved 7 February 2020.

1257) Woman in Barcelona tests positive to coronavirus, the first confirmed case on mainland Spain . The Local.ES. Retrieved 25 February 2020.

1258) Woman in Helsinki tests positive for novel coronavirus . Yle.fi. 26 February 2020. Retrieved 26 February 2020.

1259) Wong, Kai Yi (3 February 2020). Coronavirus keeps exhibitors away from Singapore Airshow; Korea's Black Eagles not coming . The Straits Times. Retrieved 24 February 2020.

1260) Wong, Kai Yi (4 February 2020). Singapore Airshow to go ahead despite fewer exhibitors, with stepped-up coronavirus measures . The Straits Times. Retrieved 24 February 2020.

1261) Wongcha-um, Panu; Fernandez, Clarence (29 February 2020). New coronavirus infection in Thailand takes tally to 42 . Reuters. Retrieved 29 February 2020.

1262) Woo PC, Lau SK, Lam CS, Lau CC, Tsang AK, Lau JH, et al. (April 2012). Discovery of seven novel Mammalian and avian coronaviruses in the genus deltacoronavirus supports bat coronaviruses as the gene source of alphacoronavirus and betacoronavirus and avian coronaviruses as the gene source of gammacoronavirus and deltacoronavirus . Journal of Virology. 86(7): 3995–4008. doi:10.1128/JVI.06540-11.

1263) Woodley, Matt (22 February 2020). Seven new coronavirus cases in Australia from quarantined cruise ship . News GP. Retrieved 23 February 2020.

1264) Woods, Amanda (19 February 2020). Diamond Princess reports 79 more coronavirus cases as expert slams 'completely inadequate' quarantine . New York Post. Retrieved 19 February 2020.

1265) Woodyatt, Amy; Kottasová, Ivana; Griffiths, James; Regan, Helen. China changed how it counts coronavirus cases again. Here's why . CNN.

1266) Woolfolk, John; Krieger, Lisa M. (28 February 2020). Second U.S. coronavirus case of unknown origin confirmed in Santa Clara County; third reported in Oregon . The Mercury News. Retrieved 28 February 2020.

1267) World Health Organization (19 February 2020). Coronavirus disease 2019 (COVID-19): situation report, 29 . World Health Organization (WHO). hdl:10665/331118.

1268) World Health Organization (28 January 2020). Clinical management of severe acute respiratory infection when novel coronavirus (2019-nCoV) infection is suspected (PDF).

1269) World Health Organization (March 2020). Coronavirus disease 2019 (COVID-19): situation report, 47 . World Health Organization(WHO). hdl:10665/331444.

1270) World Health Organization action in Afghanistan aims to control debilitating leishmaniasis . Archived from the original on 12 October 2017. Retrieved 16 November 2014.

1271) World Health Organization declares China's coronavirus a 'global health emergency' less than week after deciding not to . news.yahoo.com. Retrieved 30 January 2020.

1272) World Health Organization. Advice for Public . Archived from the original on 26 January 2020. Retrieved 10 February 2020.

1273) World Health Organization. 11 February 2020. Retrieved 24 February 2020.

1274) Worst cholera outbreak in Angola Archived 2017-04-29 at the Wayback Machine, BBC

1275) Wuhan party official apologises for failures in coronavirus outbreak . South China Morning Post. 6 February 2020. Archived from the original on 7 February 2020. Retrieved 8 February 2020.

1276) Wuhan Sharply Tightens Quarantines on Residents: Virus Update . Bloomberg. Retrieved 15 February 2020.

1277) Wuhan suspends visa, passports services for Chinese citizens until January 30: government statement . Reuters. 27 January 2020. Archived from the original on 27 January 2020. Retrieved 27 January 2020.
1278) Wuhan virus: China imposes partial lockdown in Shantou, first city to face measure outside virus epicentre . straitstimes.com. 26 January 2020. Archived from the original on 26 January 2020. Retrieved 26 January 2020.
1279) Wuhan virus: China sacks senior city health official unable to answer basic questions . The Straits Times. 30 January 2020. Archived from the original on 1 February 2020. Retrieved 8 February 2020.
1280) Wuhan virus: K-pop boy band Got7's Singapore concert on Feb 22 postponed . The Straits Times. 31 January 2020. Retrieved 24 February 2020.
1281) Wuhan virus: Temperature screening begins at Woodlands, Tuas and sea checkpoints . CNA. 24 January 2020. Retrieved 29 January 2020.
1282) www.bjnews.com.cn. Archived from the original on 30 January 2020. Retrieved 1 February 2020.
1283) www.nbd.com.cn. Archived from the original on 1 February 2020. Retrieved 1 February 2020.
1284) www3.nhk.or.jp. NHK News Web. 12 February 2020. Retrieved 12 February 2020.
1285) Xan Rice (2007-02-22). Fatal outbreak not a cholera epidemic, insists Ethiopia . The Guardian. Archived from the original on 26 October 2017. Retrieved 16 November 2014.
1286) Xia, Jianhua; Tong, Jianping; Liu, Mengyun; Shen, Ye; Guo, Dongyu (26 February 2020). Evaluation of coronavirus in tears and conjunctival secretions of patients with SARS-CoV-2 infection . Journal of Medical Virology. doi:10.1002/jmv.25725.
1287) Xiang YT, Yang Y, Li W, Zhang L, Zhang Q, Cheung T, et al. (March 2020). Timely mental health care for the 2019 novel coronavirus outbreak is urgently needed . The Lancet. Psychiatry. 7(3): 228–29. doi:10.1016/S2215-0366(20)30046-8.

1288) Xinhua (10 February 2020). Xi inspects novel coronavirus prevention, control work in Beijing . Global Times. Global Times. Retrieved 10 February 2020.

1289) xinhuanet.com. 31 January 2020. Retrieved 31 January 2020.

1290) Xu H, Zhong L, Deng J, Peng J, Dan H, Zeng X, et al. (February 2020). High expression of ACE2 receptor of 2019-nCoV on the epithelial cells of oral mucosa . International Journal of Oral Science. 12 (1): 8. doi:10.1038/s41368-020-0074-x.

1291) Xu L, Liu J, Lu M, Yang D, Zheng X (March 2020). Liver injury during highly pathogenic human coronavirus infections . Liver International. doi:10.1111/liv.14435.

1292) Ya llegó el coronavirus; México confirma tres casos; hay otros tres sospechosos (in Spanish) . Excelsior. 29 February 2020. Retrieved 29 February 2020.

1293) Yamaguchi, Mari (23 February 2020). Japan reports 3rd cruise ship death, 57 more infected . ABC News (US). Retrieved 24 February 2020.

1294) Yamaguchi, Mari (27 February 2020). Japan to Close Schools Nationwide . The Seattle Times. Retrieved 27 February 2020.

1295) Yang, Joyu Wang and Jing (4 February 2020). Out of Luck: Coronavirus Shuts Casinos in the World's Biggest Gambling Hub . Archived from the original on 6 February 2020. Retrieved 8 February 2020 – via www.wsj.com.

1296) Yanping Z, et al. (The Novel Coronavirus Pneumonia Emergency Response Epidemiology Team) (17 February 2020). The Epidemiological Characteristics of an Outbreak of 2019 Novel Coronavirus Diseases (COVID-19)—China, 2020 . China CDC Weekly. Chinese Center for Disease Control and Prevention. 2 (8): 113–122. Retrieved 18 March 2020.

1297) Yao X, Ye F, Zhang M, Cui C, Huang B, Niu P, et al. (March 2020). In Vitro Antiviral Activity and Projection of Optimized Dosing Design of Hydroxychloroquine for the Treatment of Severe Acute Respiratory Syndrome Coronavirus 2 (SARS-CoV-2) . Clinical Infectious Diseases. doi:10.1093/cid/ciaa237.

1298) Yao X, Ye F, Zhang M, Cui C, Huang B, Niu P, Liu X, Zhao L, Dong E, Song C, Zhan S, Lu R, Li H, Tan W, Liu D (March 2020). In

Vitro Antiviral Activity and Projection of Optimized Dosing Design of Hydroxychloroquine for the Treatment of Severe Acute Respiratory Syndrome Coronavirus 2 (SARS-CoV-2) . Clin. Infect. Dis. doi:10.1093/cid/ciaa237.

1299) Yasharoff, Hannah. Coronavirus: Princess cruise to stay under quarantine after 10 more passengers test positive . USA TODAY. Archived from the original on 7 February 2020. Retrieved 8 February 2020.

1300) Yeh, Joseph (19 February 2020). WUHAN VIRUS/One more confirmed coronavirus case reported in Taiwan . Focus Taiwan. Retrieved 19 February 2020.

1301) Yellow fever – countries with dengue: alert 2016-03-28 20:39:56 Archive Number: Archive Number: 20160328.4123983 . ProMED-mail. International Society for Infectious Diseases. Archived from the original on 10 April 2016. Retrieved 29 March 2016.

1302) Yong, Chean Yeah; Ong, Hui Kian; Yeap, Swee Keong; Ho, Kok Lian; Tan, Wen Siang (2019). Recent Advances in the Vaccine Development Against Middle East Respiratory Syndrome-Coronavirus . Frontiers in Microbiology. 10: 1781. doi:10.3389/fmicb.2019.01781.

1303) Young, Ian (21 February 2020). New Canadian coronavirus patient never went to China, but had visited Iran . South China Morning Post. Retrieved 21 February 2020.

1304) Ytterligere to personer har testet positivt på koronavirus (in Norwegian) . Folkehelseinstituttet. 28 February 2020. Retrieved 28 February 2020.

1305) Yuen, Sin (29 February 2020). Coronavirus: 4 new cases confirmed, all linked to Wizlearn Technologies cluster; number of S'pore cases crosses 100 . The Straits Times. Retrieved 29 February 2020.

1306) Yuill, Thomas M.; Woodall, John P.; Baekeland, Susan (2013). Latest outbreak news from ProMED-mail. Yellow fever outbreak—Darfur Sudan and Chad . International Journal of Infectious Diseases. 17 (7): e476–e478. doi:10.1016/j.ijid.2013.03.009.

1307) Yusof, Amir (10 February 2020). Malaysia confirms sixth citizen case of coronavirus, man worked in Macau before coming home unwell . CNA. Retrieved 10 February 2020.

1308) Yusof, Amir. Novel coronavirus: Malaysia confirms 1 more case, Wuhan woman was in Singapore with family . CNA. Archivedfrom the original on 7 February 2020. Retrieved 7 February 2020.

1309) Yusor, Teh Athira (27 January 2020). Malaysia suspends visas for Chinese tourists from Wuhan and Hubei . New Straits Times. Retrieved 28 January 2020.

1310) Zabdiel Boylston and inoculation. Archived from the original on 5 September 2015. Retrieved 16 November 2014.

1311) Zack, Justin. Coronavirus: 16th patient to test positive is woman from Wuhan (Updated) . The Star. Retrieved 8 February 2020.

1312) Zeitung, Süddeutsche. Coronavirus: Flug in Frankfurt gelandet . Süddeutsche.de (in German). Archived from the original on 2 February 2020. Retrieved 7 February 2020.

1313) Zeitung, Süddeutsche. Evakuierungsflug aus China nach Umweg gelandet . Süddeutsche.de (in German). Archived from the original on 2 February 2020. Retrieved 7 February 2020.

1314) Zhang H, Penninger JM, Li Y, Zhong N, Slutsky AS (March 2020).

1315) Zhang, Wei; Du, Rong-Hui; Li, Bei; Zheng, Xiao-Shuang; Yang, Xing-Lou; Hu, Ben; Wang, Yan-Yi; Xiao, Geng-Fu; Yan, Bing; Shi, Zheng-Li; Zhou, Peng (1 January 2020). Molecular and serological investigation of 2019-nCoV infected patients: implication of multiple shedding routes . Emerging Microbes & Infections. 9 (1): 386–389. doi:10.1080/22221751.2020.1729071. PMC 7048229.

1316) Zheng YY, Ma YT, Zhang JY, Xie X (March 2020). COVID-19 and the cardiovascular system . Nature Reviews. Cardiology. doi:10.1038/s41569-020-0360-5.

1317) Zhou F, Yu T, Du R, Fan G, Liu Y, Liu Z, et al. (2020). Clinical course and risk factors for mortality of adult inpatients with COVID-19 in Wuhan, China: a retrospective cohort study . The Lancet. Elsevier BV. doi:10.1016/s0140-6736(20)30566-3. ISSN 0140-6736.

1318) Zhou P, Fan H, Lan T, Yang XL, Shi WF, Zhang W, et al. (April 2018). Fatal swine acute diarrhoea syndrome caused by an HKU2-related coronavirus of bat origin . Nature. 556 (7700): 255–58. Bibcode:2018Natur.556..255Z. doi:10.1038/s41586-018-0010-9.

1319) Zhou P, Yang XL, Wang XG, Hu B, Zhang L, Zhang W, et al. (23 January 2020). Discovery of a novel coronavirus associated with the recent pneumonia outbreak in humans and its potential bat origin . bioRxiv (preprint). doi:10.1101/2020.01.22.914952.

1320) Zhuo, Tee (1 February 2020). Singapore confirms 2 more Wuhan virus cases, including 2nd Singaporean woman on Scoot flight from Wuhan . straitstimes.com. The Straits Times. Archived from the original on 1 February 2020. Retrieved 1 February 2020.

1321) Ziady, Hanna (26 February 2020). Biotech company Moderna says its coronavirus vaccine is ready for first tests . CNN. Archived from the original on 28 February 2020. Retrieved 2 March 2020.

1322) Zwei Fälle in Tirol bestätigt (in German). ORF. 25 February 2020.

****

yes

I want morebooks!

Buy your books fast and straightforward online - at one of world's fastest growing online book stores! Environmentally sound due to Print-on-Demand technologies.

Buy your books online at
www.morebooks.shop

Kaufen Sie Ihre Bücher schnell und unkompliziert online – auf einer der am schnellsten wachsenden Buchhandelsplattformen weltweit! Dank Print-On-Demand umwelt- und ressourcenschonend produzi ert.

Bücher schneller online kaufen
www.morebooks.shop

KS OmniScriptum Publishing
Brivibas gatve 197
LV-1039 Riga, Latvia
Telefax: +371 686 204 55

info@omniscriptum.com
www.omniscriptum.com

Printed by Books on Demand GmbH, Norderstedt / Germany